Nanovate

Mohab Anis • Ghada AlTaher • Wesam Sarhan
Mona Elsemary

Nanovate

Commercializing Disruptive Nanotechnologies

Springer

Mohab Anis
American University in Cairo
New Cairo, Egypt

Ghada AlTaher
American University in Cairo
New Cairo, Egypt

Wesam Sarhan
American University in Cairo
New Cairo, Egypt

Mona Elsemary
American University in Cairo
New Cairo, Egypt

ISBN 978-3-319-44861-9 ISBN 978-3-319-44863-3 (eBook)
DOI 10.1007/978-3-319-44863-3

Library of Congress Control Number: 2016960538

Printed on acid-free paper

This Springer imprint is published by Springer Nature
The registered company is Springer International Publishing AG
The registered company address is: Gewerbestrasse 11, 6330 Cham, Switzerland

Acknowledgements

The authors would like to acknowledge the help of all the people involved in this project. We would like to acknowledge the efforts of Dr. Hanadi Salem and Dr. Wael Mamdouh for reviewing the book, as well as the efforts of Mariam Elsherbiny who edited the book in terms of structure. Finally, the support we received from Springer's editorial team was outstanding. Without their support, this book would not have become a reality.

The business community usually perceives nanotechnology as a complex field of science. What this book does is demonstrate how nanotechnology can be applied to different sectors to make products more competitive and manufacturing processes more efficient. This book was inspired by the many stories, applications and case studies that emerged from the postgraduate course taught at the American University in Cairo, Economics and Management of Nanotechnology. We are indebted to the contributions made by postgraduate students who came from all disciplines of nanotechnology.

Disclaimer from Authors

Readers should be aware that the internet websites or the cited reports listed in this work may have changed or disappeared between when this work was written and when it is read. The presented information was based on analysis and predictions by the authors, and not the companies or their representatives.

Contents

About the Authors

Ghada AlTaher holds a master's degree in Nanotechnology from the American University in Cairo. She is passionate about the emerging new trends in business and technology innovations. Currently, she works in the Egyptian Drug Authority (EDA), where she helps pharma companies abide by the current regulations that govern the registration and marketing of human drugs in the Egyptian market. Prior to EDA, she used to work as a teaching assistant in the German University in Cairo. She holds a Bachelor of Science degree in pharmacy and biotechnology.

Mohab Anis is an engineering professor at the American University in Cairo and CEO at INNOVETY, a leading innovation management consultancy. Mohab loves helping companies build new business concepts and has consulted widely in the areas of innovation strategy, capacity and discipline. This includes auditing businesses' innovation capabilities, idea generation and management, and technology forecasting. Mohab has also worked with government on strategy and policy that fosters national innovation and entrepreneurship. Mohab is a certified consultant of A. T. Kearney's House of Innovation and has delivered numerous keynotes and workshops in the area of innovation management.

Earlier, Mohab spent years as a tenured professor of Computer Engineering at the University of Waterloo, Canada, to which he's now adjunct. He consulted in the USA and Canada to a variety of Fortune 500 companies. As an academic, he authored over 150 international papers and 5 books. He is on the editorial board of 9 international journals and has supervised over 30 postgraduate students, half of which were Ph.Ds. Mohab was awarded three of Canada's highest awards for excellence in innovation (The Early Research Award from Ontario's Ministry of Research and Innovation, the Colton Medal for Research Excellence and the IEEE International Design Award), as well as both of AUC's top awards in Teaching and Research & Creative Endeavours.

Mohab holds a Ph.D. in Computer Engineering from the University of Waterloo (2002), an M.B.A. from Wilfrid Laurier University with a concentration in Innovation and Entrepreneurship (2008), and a Master's in Management Sciences with a concentration in Technological Innovation (2008).

Mona Tarek Elsemary is a Yousef Jameel Ph.D. Fellow and teaching assistant at the American University of Cairo/Egypt. She has earned two master's degrees: one in Microbiology from Cairo University/Faculty of Pharmacy in Cairo, Egypt, and another in Pharmaceutical Biotechnology from King Saud University/Faculty of Pharmacy in Riyadh, Saudi Arabia. She has published two papers, one of which was presented at the San Francisco Conference for Pharmaceutic and Novel Drug Delivery Systems in 2012. She has expertise in different drug formulation methods, electrospinning and preparation of nanofibre as well as bacterial biofilm targeting with novel drug delivery methods.

Wessam Sarhan is the co-founder and CEO of NanoGreen Institute, which represents Egypt's' first platform for green nanotechnology education and research serving students, researchers, industry and communities. Wessam is also the co-founder and technology consultant of NanoEbers Startup Company for nano-based innovative wound dressings.

Chapter 1
Introduction

Introductory Words

"*There is plenty of room at the bottom*"

In 1960, the famous physicist Richard Feynman gave a famous lecture titled *There is Plenty of Room at the Bottom*. He addressed the potential of miniaturization in developing computer and Information Technology systems. That lecture is considered to be the point of origin which predicted the evolution of a new science: Nanotechnology.

Nanotechnology (nano) is the science that deals with the properties of materials at the nano-scale. 1 nanometer (nm) is equal to 10^{-9} m. To better imagine how small this value is, consider that the relationship between 1 nm and 1 m is—relatively—the same relationship between a soccer ball and Earth!

How do we create nanotechnology?

Bottom-Up and Top-Down, are the two approaches that are commonly used in the synthesis of nanomaterials. To make it simple, consider that you are making a wooden doll. In the Bottom-Up approach, the doll would be fabricated from small pieces of wood assembled together through an optimized process. However, in the Top-Down approach, a large piece of wood would be crafted into the form of that wooden doll. Usually the Top-Down approach requires sophisticated techniques and equipment such as lithographic techniques and molecular beam epitaxy. The good news is that some of the Bottom-Up techniques can proceed via simple chemical synthesis procedures such as the synthesis of silver nanoparticles or gold nanoparticles. However, in all cases, strong analytical tools like Atomic Force Microscopy and Electron Microscopy are essentially required.

Why is nanotechnology interesting?

Materials at this scale exhibit enhanced properties compared to the same materials in the bulk (micrometers to meters in size).

M. Anis et al., *Nanovate*, DOI 10.1007/978-3-319-44863-3_1

In addition, they sometimes behave adversely to their bulk counterparts. For example, consider gold nanoparticles which have become a benchmark for nanotechnology in the past few years and have become highly popular in many applications. At this scale they are highly reactive and can be used to trigger reactions. On the other hand, bulk gold particles are completely inert and rarely oxidize or change in response to their surroundings—that's why they are widely used in jewelry. Through the use of nanotechnology, materials can be made more effective within the intended application. Adopting nanotechnology can result in more durable, lighter, more reactive, more porous or less porous, more conductive or less conductive materials according to the aim of the implementation.

Another very important feature of nanotechnology that has led to its strong growth is that the developed product does not need to be nano-related to utilize the benefits nanotechnology offers. In other words, adding a nano-feature or a nanomaterial in a new or existing product could enhance its properties drastically. This has been observed in different products and in different fields ranging from medicine, electronics, and automotive to textiles, food and cosmetics. Nanotechnology does not only have the potential to boost businesses, it also plays an important role in addressing and solving challenges that the global world is facing: shortages of raw materials or resources, clean energy problems, fuel depletion, bacterial resistance and cancer.

A bright opportunity

According to the 2014 Business Communication Company market research report, the global market for nanotechnology products was valued at 22.9 billion dollars in 2013 and increased to about 26 billion dollars in 2014. It is expected that the nanotechnology market will reach about 64.2 billion dollars by 2019, at a Compound Annual Growth Rate (CAGR) of 19.8% from 2014 to 2019 [1]. Currently, nanotechnology research and development is widely spread. Numerous universities and institutes are engaged in active nanotechnology research in various fields such as nanomedicine, nanoelectronics, engineering and energy. Moreover, governments have realized the necessity of nanotechnology research; in 2001, the Bush administration increased funding for nanoscale science by 16%. Thus, nanotechnology research is viewed as an opportunity with the potential of bridging gaps between industries and impacting our daily lives.

An eye opener

It is important to realize that nanotechnology innovations should not be regarded as standalone products; nano-based products are just translations for: a nano-based property utilized into a particular tool to serve a particular need. Consequently, the uniqueness of nanotechnology innovations is that the same technology or nanostructure can be tailored to serve different applications in many industrial sectors. This leads us to say that innovators in nanotechnology are those who keep their eyes open on the advancements in nanotechnology in different sectors. For example, Saudi Aramco spotted an opportunity of developing nanobots (nano-robots) for oil reservoirs from the healthcare sector. In that sector, scientists were working on developing nanobots that could travel through the human circulatory system to deliver drugs or detect diseases. Saudi Aramco noticed that oil is stored inside rocks in tiny vessels that form a network mimicking that of the human circulatory system.

Disruptive Technology

Nanotechnology innovations can overturn markets because the new products are more efficient and/or cheaper than existing products that once lead the market. Therefore, industries which do not adapt to these revolutionary innovations are likely to fail and disappear.

How to convince customers to buy what they can't see

Public perception of nanotechnology is positive, yet still not mature enough because of the of lack adequate knowledge about the potential benefits associated with this new technology. This can be attributed to that the public has not yet recovered from the hazardous failure of the technology of "Genetic Engineering". Since the public still has an inherent fear of new technologies, companies should only use the word *nano* with caution when marketing their products. Companies should make sure their target markets understand the benefits associated with the technology and that the word "nano" affects purchasing decisions positively. In many other cases, the marketing strategy for nano-based products does not reveal that there is a nano-contribution to a particular product. One of the main marketing arguments for this is that the focus usually is not on the nanotechnology associated with that product.

Types of Nanostructures

This section provides an overview of the types of nanostructures that will be covered in the book.

Graphene

Graphene is a two-dimensional-planar single atomic layer crystal structure formed of hexagonally arranged sp2 bonded carbon atoms in this single layer. It is an allotrope of carbon and is a structural element of other allotropes like carbon nanotubes (cylindrically rolled up graphene), graphite (stacked layers of graphene) and fullerene (graphene folded into a sphere). This is shown in Fig. 1.1. Seventy years ago, scientists would have argued that single layer two-dimensional crystals could never exist. Due to thermodynamic instability and that these single layers would have very low melting points, the atoms would rearrange to form clusters or islands or decompose [2, 3]. Up until 2004, two-dimensional materials were only thought to be parts of

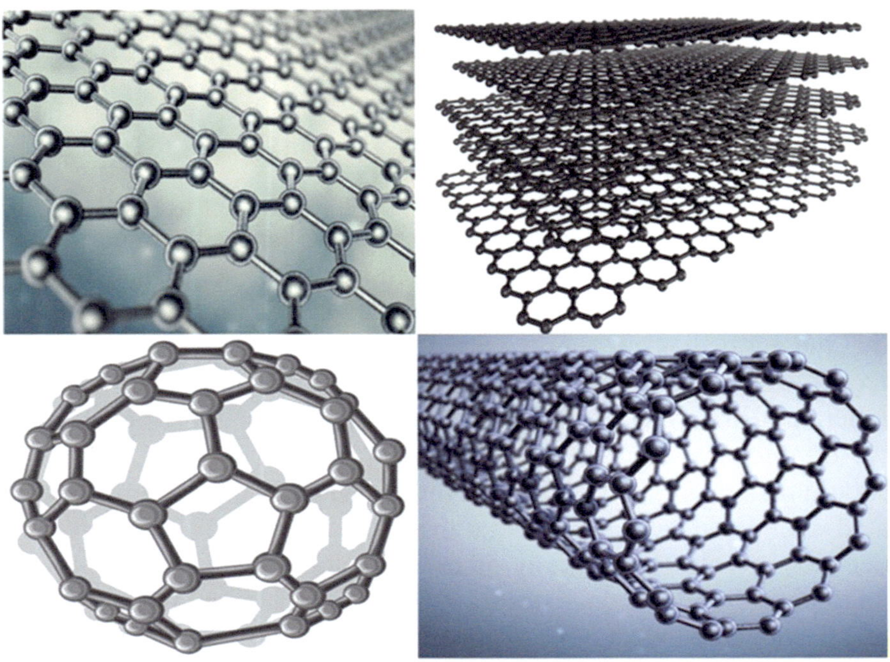

Fig. 1.1 Structures of graphene, graphite, fullerene and single walled carbon nanotube compared

three-dimensional structures and that they could not exist alone [4]. In 2004, however, graphene was discovered along with other two-dimensional crystals like boron-nitride single layer. These single layers were formed in liquid suspensions as membranes and were found to exhibit high qualities. Such enhanced qualities include improved mechanical properties as well as, in the case of graphene, very high electrical conductivity [5–7].

This single layer crystal is formed of carbon atoms arranged in a hexagonal lattice. Each carbon atom is bonded to three carbon atoms separated by a distance of 1.42 Å. Each carbon atom is bonded with three σ bonds and one out of plane π electrons responsible for the very good electrical conductivity. This sp2 hybridization and close packing of the carbon atoms make the graphene stable. If holes are formed in the sheet of graphene it can easily selfheal if bombarded or exposed to pure carbon atoms or carbon atoms in a molecule [8]. Graphene sheets, however, tend to cluster and layer in solid form as they crystallize from a melt, and show (002) layering in XRD. Unlayered solidified graphene was only found in presolar graphite onions cores [9]. As a result, graphene typically occurs as either monolayers attached to a substrate surface of metal or SiC in solid form or as suspended free standing graphene [10].

Graphene's high electron mobility and very low resistance makes it an ideal candidate in electronic applications [11] in addition to being transparent, mechanically strong and flexible which makes it useful in the fabrication of transparent conducting films. Graphene has been used in applications such as e-paper: organic light-emitting diodes and touch screens [12]. Moreover, graphene has special optical properties and is capable of absorbing energy at the wide band microwave to ultraviolet signals. This makes it ideal for use in photo-modulators, laser transmission, terahertz wave detectors and photodetectors [13]. It has also been widely used in sensors, supercapacitors and in battery electrodes [14].

Graphene can be synthesized by mechanical, chemical and thermal exfoliation as well as other more complex techniques as vapor deposition techniques and arc discharge techniques using an electric current. One method, mechanical exfoliation, is accomplished with the aid of the Atomic Force Microscope (AFM) and the Scanning Tunneling Microscope (STM) that can be used to exfoliate single crystals of graphite and Highly Ordered Pyrolytic Graphite (HOPG). Chemical exfoliation on the other hand involves the aid of a chemical intercalating agent in a two-step process. A certain agent is squeezed between the graphite layers to reduce the forces between them and then, through sonication or heating, they are separated. In vapor deposition techniques, a hydrocarbon gas (such as methane) is broken down and deposited at a high temperature and pressure to form graphene sheets on a metal substrate (Ni, Pd, Cu, etc.). In the arc discharge method, an extremely high voltage is used to evaporate graphite at a high pressure of hydrogen gas to deposit thin layers of graphene. Different fabrication techniques and modifications of such techniques have been researched and carried out to form functionalized graphene, graphene composites with metals, ceramics and polymers to be used to serve different applications.

Carbon Nanotubes

Carbon nanotubes are cylindrical allotropes of carbon which can be viewed as sheets of graphene rolled up into a cylinder. They can be single walled (made of a single cylinder), double walled (made of two concentric cylinders), triple or multi-walled. Each one of these will have different properties and thence different applications. In addition, the cylinder diameter and length will also affect the properties and function. The direction through which the cylinders are rolled or the chirality is also a necessary determinant in the properties of the carbon nanotubes. Moreover, nanotubes can be bent in a donut shape called nanotorus with superior thermal stability and magnetic properties to nanotubes [15]. Carbon nanotubes can also be attached to other carbon allotropes or doped with metals or other materials to modify their properties and then be customized for their suitable applications. Carbon nanotubes can be attached to fullerenes (sphere shaped allotrope of carbon) to form nanobuds. These nanobuds have combined properties of both nanotubes and fullerenes and serve to prevent the slipping or bending of the carbon nanotubes when dispersed in composites thence anchoring them better in the composite to produce better mechanical properties. Carbon nanotubes can also be attached to graphene sheets aligned along the sidewalls of the carbon nanotubes. They yielded excellent supercapacitor functions due to the large reactive surface area formed [16]. Furthermore, there are carbon nanotubes with fullerene inside them that have modified magnetic properties with heating [17]. Nitrogen doping of carbon nanotubes is carried out to enhance storage capacity in batteries [18, 19]. There are also functionalized carbon nanotubes that are used in sensors. Whether carbon nanotubes are used alone or modified, they have proven to be highly effective in their applications due to their outstanding properties.

Carbon nanotube, until now, is the strongest, stiffest and hardest material discovered on Earth. It has a switchable hydrophilicity so when no voltage is applied it is highly hydrophobic and when a voltage is applied it becomes highly hydrophilic [20]. Carbon nanotubes have interesting electrical properties. They can be superconductors or semiconductors affected by the chirality of the nanotube [21]. Another interesting property seen in multi-walled carbon nanotubes are that the tubes slide inside each other with zero friction and can rotate relative to one another also with zero friction which has aided research in designing the first molecular motor.

The synthesis of carbon nanotubes is also done by similar processes as graphene but with different conditions as temperature, pressure, catalyst and starting materials. They are also synthesized by chemical vapor deposition and arc discharge. Carbon nanotubes are used for a variety of applications which can be affected by the synthesis mechanisms. A limiting property of carbon nanotubes is that they tend to aggregate due to the high surface energy. Such aggregation compromises their beneficial properties to some extent, however, they are still found superior in their applications despite that.

Currently, carbon nanotubes are used to strengthen composites in applications such as sports' equipment like bike frames, boats and vessels. They are also used to develop high storage batteries, supercapacitors, transistors and electronic circuits. Additionally, they are used to make small, transparent circuits and in the near future will be used to make flexible screens and electronic paper. They are used in solar cells and hydrogen storage for fuel cells as well. Carbon nanotubes are used to make nanosponges that are capable of binding and removing contaminants from water and air and can be used for environment remediation [22].

Fullerene

Fullerene is an allotrope of carbon, comprised of carbon atoms arranged in a spherical shape. It can be formed of 20 or 60 carbon atoms. The carbon atoms can be arranged in pentagon or hexagon rings to form the spherical ball. Fullerenes are synthesized by vaporization of graphite or another carbon source usually by employing a by pyrolysis, arc discharge-plasma, or radio-frequency-plasma techniques. There is increasing research now in the use of chemical synthesis approaches to produce fullerenes but haven't yet been commercially used [23]. Fullerenes have numerous applications that are being discovered more every day. It is a greatly reactive molecule when functionalized and can be used as a scavenger for reactive oxygen species and free radicals. This has made them useful as antioxidants for use in the health and cosmetics sectors. They have also been found to be useful in the slowing down of nerve damage associated with diseases like Alzheimer's.

Nanopolymers (Conductive)

Polymers are long chains of repeating units called monomers. When any one dimension of the polymer is in the nano range, it is called a nanopolymer. They can be one dimensional e.g. a nano thin homogeneous polymeric film, two dimensional e.g. nanotubes, nanoconduits, nanofibres, nanostructures arranged on a polymer surface, three dimensional e.g. nanocapsules, nanospheres, dendrimers, self-assembled structures, hyper-branched polymers, porous materials or a combination of all three to form composites. Such nanopolymers can have a self-assembly nature or none. Examples of commonly used polymers include polycaprolactone, poly lactide glycolide, polystyrene, polyvinylpyrrolidone, chitosan, dextran and proteins. Polymers can either exhibit elastic, plastic or brittle mechanical properties. Conductive polymers are a new class of polymers that has recently been of increasing in interest. These are organic compounds that can conduct electricity; they can act as metal conductors or as semiconductors. Examples include polyaniline, polypyrrole, polyacetylene and many others. Conductive polymers have a wide range of applications namely in electronic circuits, solar cells, capacitors, light emitting diodes, et cetera. Polymers, in general, are very versatile in applications, involved in nearly every industry including packaging, making composites for mechanical components, making artificial limbs and tissue engineering.

Nanoparticles

Nanoparticles represent a major class of nanomaterials. They are zero-dimensional solid microscopic particles that are between 10 and 1000 nm in size [24]. Nanoparticles are of great interest to different fields like the electronic, energy, information, environmental and medical fields. Metals, metal oxides and semiconductors are more involved in the energy, information and electronic fields [25], while polymeric nanoparticles are more involved in the food, cosmetic and medical fields. At the nanoscale, particles exhibit unexpected properties in addition to increased surface to volume ratio. Thus, one of the goals for designing and using a nanoparticle in consumer products would be to make use of the new properties achieved at the nanoscale as in the case of silver and titanium oxide nanoparticles (Fig. 1.2).

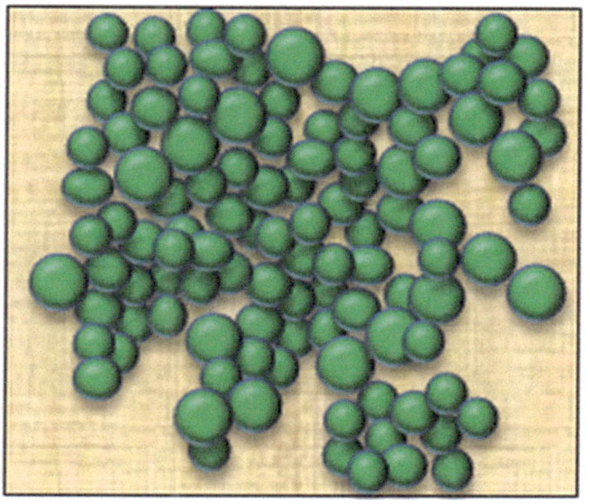

Fig. 1.2 Illustrative figure demonstrating nanoparticles of different diameter distribution

Silver nanoparticles, for example, exhibit antimicrobial properties at the nanoscale, thus, they are being extensively used in cosmetics, drugs, coatings and packaging materials. Another major goal for the use of nanoparticles would be as a delivery system; making use of the increased surface to volume ratio of the nanoparticles and the increased ability for penetration, distribution and bioavailability. In designing a nanoparticle as a delivery system, the main factors to be controlled would be the size of the synthesized particles and their surface properties as well as the ability of the particles to release their load at the specific and required site in a controlled and sustained fashion [24]. The use of nanoparticles as a delivery system involves a number of advantages including the ability to easily manipulate particle size and surface characteristics to achieve specific and optimum delivery of the carried load and ability to control the release profile of the loaded material as well as the degradation rate of the nanoparticle. However, nanoparticles suffer from a number of limitations that need to be overcome to facilitate their commercial availability. Such limitations include burst release of the carried load in addition to particle aggregation that complicates their physical handling either in the liquid or the solid form.

Nanoparticles can be assembled via a variety of methods some of them very simple and others more complicated. Polymeric nanoparticles can be fabricated via dispersion of preformed polymers which can be done via the solvent evaporation method, spontaneous emulsification/solvent diffusion, nano-precipitation, salting out, dialysis and supercritical fluid technology. Other methods for fabrication of polymeric nanoparticles include polymerization of monomers, ionic gelation or coacervation, supercritical fluid technology as well as Particle Replication in Nonwetting Templates (PRINT) which have been observed to have complete control over the particle shape and size [24]. Fabrication of metallic nanoparticles on the other hand could be achieved via chemical reduction, sol-gel, hot-soap, reversed micelle, spray pyrolysis and other pyrolysis methods. Chemical reduction, which is a simple technique, has been extensively and successfully used for the fabrication of Noble metal nanoparticles [25].

Nanoparticles can be further classified, according to their composition, into nanocapsules, nanomicelles, nanoemulsions, nanospheres and nanocrystals. Such classes of nanomaterials will be briefly illustrated in the following sections.

Nanomicelles

Nanomicelles are colloidal dispersions of self-assembling particles with dimensions ranging from 10 to 100 nm. Nanomicelles are characterized by having a hydrophobic core and a hydrophilic shell [26]. Nanomicelles are specifically advantageous as a delivery system for hydrophobic materials via solubilizing the hydrophobic material and entrapping it within the micellar hydrophobic core. At the same time, hydrophilic chains extend outwards resulting in a readily soluble aqueous formulation that can be easily administered via different administration routes [27]. Nanomicelles are further characterized by their small size in addition to their low toxicity. However, further research is still needed to optimize the sustained release from the small size micelles as compared to other larger particulate systems as liposomes and nanoparticles [28].

Nanocapsules

Nanocapsules are particles composed of a liquid core—mostly an oil core—that is surrounded by a dense polymeric matrix. This advantageous oily core allows the delivery of hydrophobic materials efficiently. They can be fabricated via two techniques namely interfacial polymerization and the emulsification diffusion techniques. Nanocapsules aid in preventing drug precipitation during preparation and the subsequent stability problems that result from the incorporation of the drug in the core compared to other nano-based delivery systems where the drug is loaded on the surface [29].

Nanoemulsions

Nanoemulsions are colloidal dispersions of nano-scale droplets that are formed via shear-induced rupturing. They consist of a lipid phase that is dispersed in an aqueous continuous phase. In nanoemulsions, every oil droplet is surrounded by a thin interfacial layer of emulsifying molecules [30]. Due to their small sizes, nanoemulsions show good stability against aggregation as well as gravitational separation. Nanoemulsion-based delivery systems also show the advantage of increasing the bioavailability of the encapsulated material. Different methods have been developed for the fabrication of the nanoemulsions and include solvent displacement, precipitation, emulsification diffusion and emulsification evaporation [31]. The nanoemulsion fabrication techniques can be categorized into high-energy methods and low energy methods. Although the high energy methods offer enhanced control on the size of the developed nanoemulsions, they are extremely strenuous to scale up, unlike the low energy techniques which are more feasible for scaling up [32].

Nanospheres

Nanospheres are particles that are spherical in shape and exhibit sizes within the range of 10–200 nm. Within the process of formulating nanospheres, the material to be loaded is dissolved, encapsulated, attached or entrapped to the polymer matrix. Nanospheres show the advantage of protecting the material loaded from chemical and enzymatic degradation [33]. However, due to the hydrophobic surfaces of these particles they are prone to elimination via the reticuloendothelial system. Thus, in order to prolong the circulation of the nanospheres within the human body and enhance their compatibility, their surface has to be modified to be more hydrophilic. This could be achieved via the adsorption of different surfactants on the surface of the nanospheres. Various methods are used for nanosphere fabrication and include phase inversion temperature, polymerization, solvent evaporation and solvent displacement techniques [34].

Quantum Dots (QD)

Quantum Dots (QD) are minute particles in the range of 2–10 nm. They are made of the nanocrystals of semiconductors such as cadmium selenide or lead sulfide. Moreover, they exhibit enhanced and unique electronic, magnetic and optical properties. Such uniqueness stems from the fact that these properties can be tuned by controlling their sizes and composition. At such nano sizes, quantization effects take place, i.e. electrons in QD are located in quantized energy levels which are separated by band gaps. By applying an external energy source, electrons get excited to higher energy levels, and different wavelengths (colors) are emitted depending on the quantum dots' respective sizes. As they get smaller, the band gap gets bigger, so they emit colors of higher energies. Figure 1.3 illustrates this. That's why QD plays an important role in the next generation lighting materials, diagnostics, and solar cells [35].

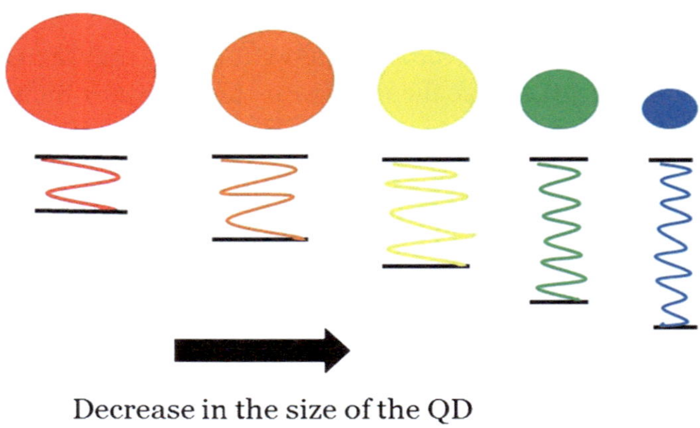

Fig. 1.3 The figure illustrates how QD emit higher energy colors as they get smaller

The benchmark and widely applicable mode of synthesis of quantum dots is using the one pot wet chemistry mode. In such a method, organometallic precursors are injected into a hot organic solvent containing surfactants (to prevent the aggregation of QD nanocrystals) at 120–360 °C under vigorous stirring. The high temperatures initiate the decomposition of the precursors, which then undergo nucleation followed by crystal growth until the required sizes of the nanoparticles are reached. Hereafter, the process is arrested by cooling. The resulting QD can be decorated by different molecules to confer a certain property, such as water solubility in biomedical applications. This mode produces high quality QD with uniform a size distribution. Their sizes can be accurately controlled by controlling the process time, temperature, and the precursors' concentration [36, 37].

huge research on how to control their properties and applications with very interesting results. Now, QD are reshaping the market landscape by the frequent launching of new advanced products especially in the fields of bio-medics, solar cells and optoelectronics such as displays for TV sets and tablets. The market size for QD is expected to reach 4704.86 million dollars by 2020 [38]. The key drivers for such expected growth is that the technology is more efficient and can be up-scaled. In addition, large corporations in the display and energy markets are eager to provide more advanced and cost effective products. Major players in the QD market globally are Nanoco Group Plc in the United Kingdom, Nanosys, Inc. and QD Vision, Inc. in the United States [38].

Nanocomposite

Composites are solid state mixtures of two or more components. They are called nanocomposites because one or more of the components are nanosized. Composites show unusual property combinations. When one or more components are on the nanoscale, these properties are tremendously enhanced, allowing even new properties to show up. The properties of nanocomposites are determined by the interfacial interactions between its components at the boundaries. Components on the nanoscale have larger surface areas, so they offer newer and larger surfaces for interactions. Consequently, these components allow for enhanced properties without too high of a usage.

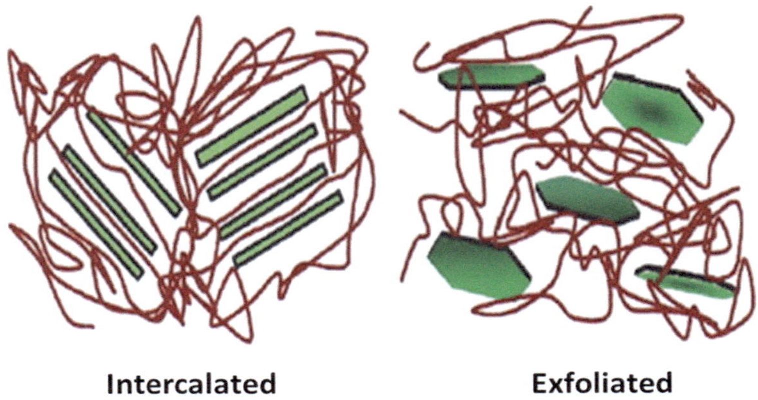

Fig. 1.4 The figure shows polymer (matrix) silicate (nanofiller) nanocomposites in which: (*1*) Intercalated: whenever the silicates are separated but have a structural arrangement with respect to each other; (*2*) Exfoliated: whenever the silicates are randomly and totally separated from each other within the matrix. Reproduced with permission [40]

Nanocomposites, structurally, are made of a matrix and a filler, with the latter dispersed in the former. The matrix could be metal, ceramic or polymer based. Fillers could be nanoparticles, nanotubes, or nanofibers. This is depicted in Fig. 1.4.

Nanocomposites' market is one of the fastest growing markets, with a continuously rising demand. The market size of nanocomposites is expected to reach 5.91 billion dollars by 2018 [39]. Because they serve nearly all types of markets, they are currently used in food packages (as barriers against moisture and gas), in automotive parts (as enhanced mechanical properties), in many households (for lighter and more durable materials), in sports equipment (for extra strength) and in construction materials (for enhanced mechanical properties).

The processing of nanocomposites begins by preparing the nanofillers. Preparation of nanofillers depends on the type of the nanomaterial. For example, Carbon Nanotubes (CNTs) are commonly prepared by Chemical Vapor Deposition or Physical Vapor Deposition methods. Generally, nanofillers are prepared from the gas phase or the liquid phase by chemical methods. In the gas phase synthesis method, the precursor is either decomposed or vaporized thermally to yield atoms that nucleate and grow in the gas phase to the desired nano size. In the liquid phase synthesis method, the starting materials react chemically in the liquid phase to produce the desired material whose size can be controlled by using surfactants.

After the preparation of the nanofillers, nanocomposites are commonly created by mixing the nanofillers and the matrix under high pressures or temperatures. Usually, nanocomposites can be created also by wet chemical modes such as the Sol-Gel technique in which precursors interact chemically in the Sol phase and polymerize, yielding an interpenetrated gel network.

Nanowires and Nanowhiskers

A human hair is around 50–120 μm wide. Imagine that you subdivided it 50,000 times. You would get one-dimensional nanostructures called nanowires (previously known as nanowhiskers). Whenever we deal with nanosizes, quantum confinement takes place. That's why nanowires are sometimes referred to as quantum wires.

Nanowires have remarkable electronic and optical properties that have revolutionized many fields especially that of Electronics and Communication. Nanowires and nanowhiskers are characterized by very high aspect ratios, meaning that they have very thin diameters (nanometers) but their lengths can extend up to micrometers. The material type from which they are constructed determines their intended application starting from stain resistant nanowhiskers in fabrics to the highly advanced and complicated configurations of nanowires in tiny transistors used in superfast computers [35]. They could be made of organics, metals like silver, or semiconductors.

Nanowires are the key role players in many technologies ranging from simple household appliances to complex smart devices in chips and computers. They have a huge surface-to-volume ratio offering large surfaces for the fast transfer and exchange of charges. Because of that, nanowires have been deeply researched and have shown promising results in displays, batteries, solar cells, and cell phones. Each industrial segment offers potential for the growth of a market through a certain type of nanowire. For example, silver nanowires are being extensively employed in Transparent Conductor (TC) materials used mainly in touch screens. The market size for silver nanowires in TC is around 41 million dollars in 2015 and is expected to reach 158 million dollars by 2018 [41]. The next generation of futuristic electronics is targeted towards the development of flexible smart devices and wearable electronics. Nanowires are the main contributors to the development of such futuristic devices.

Nanowires can be synthesized using the Top-Down approach in which a big wire is chemically etched or bombarded by highly energetic particles under vacuum to get a nanowire. Additionally, they can be synthesized by a Bottom-Up approach commonly known as vapor-liquid-solid method in which the particles in a precursor vapor are grown on a catalyst until supersaturation creating nucleation sites for the growth and deposition of one-dimensional nanostructures [42].

References

1. Nanotechnology: A Realistic Market Assessment (2014) BCC market research
2. Venables JA, Spiller GDT, Hanbucken M (1984) Nucleation and growth of thin films. Rep Prog Phys 47:399–459
3. Mermin ND (1968) Crystalline order in two dimensions. Phys Rev 176:250–254
4. Evans JW, Thiel PA, Bartelt MC (2006) Morphological evolution during epitaxial thin film growth: formation of 2D islands and 3D mounds. Surf Sci Rep 61:1–128
5. Meyer JC, Geim AK, Katsnelson MI (2007) The structure of suspended graphene sheets. Nature 446:60
6. Stankovich S, Dikin DA, Dommett GHB, Kohlhaas KM, Zimney EJ, Stach EA, Piner RD, Nguyen ST, Ruoff RS (2006) Graphene-based composite materials. Nature 442:282–286
7. Novoselov KS, Geim AK (2007) The rise of graphene. Nat Mater 6:183–191
8. Zan RR, Quentin M, Bangert U, Novoselov KS (2012) Graphene re-knits its holes. Mesoscale Nanoscale Phys 12:3936–3940

9. Bernatowicz TJ, Cowsik R, Gibbons PC, Lodders K, Fegley B Jr, Amari S, Lewis RS (1996) Constraints on stellar grain formation from presolar graphite in the Murchison meteorite. Astrophys J 472:760–782
10. Shivaraman S, Barton RA, Yu X, Alden J, Herman L, Chandrashekhar MVS, Park J, McEuen PL, Parpia JM, Craighead HG, Spencer MG (2009) Free-standing epitaxial graphene. Nano Lett 9:3100–3105
11. Schwierz F (2010) Graphene transistors. Nat Nanotechnol 5(7):487–496
12. Novoselov KS, Fal'ko VI, Colombo L, Gellert PR, Schwab MG, Kim K (2012) A roadmap for graphene. Nature 490(7419):192–200
13. Wang F, Zhang Y, Tian C, Girit C, Zettl A, Crommie M, Shen YR (2008) Gate-variable optical transitions in graphene. Science 320(5873):206–209
14. Zhu Y, Murali S, Stoller MD, Ganesh KJ, Cai W, Ferreira PJ, Pirkle A, Wallace RM, Cychosz KA, Thommes M, Su D, Stach EA, Ruoff RS (2011) Carbon-based supercapacitors produced by activation of graphene. Science 332(6037): 1537–1541
15. Huhtala M, Kuronen A, Kaski K (2002) Carbon nanotube structures: molecular dynamics simulation at realistic limit. Comput Phys Commun 146(1):30–37
16. Parker CB, Raut AS, Brown B, Stoner BR, Glass JT (2012) Three-dimensional arrays of graphenated carbon nanotubes. J Mater Res 27:1046–1053
17. Smith BW, Monthioux M, Luzzi DE (1998) Encapsulated C-60 in carbon nanotubes. Nature 396(6709):323–324
18. Shin WH, Jeong HM, Kim BG, Kang JK, Choi JW (2012) Nitrogen-doped multiwall carbon nanotubes for lithium storage with extremely high capacity. Nano Lett 12(5):2283–2288
19. Experimental nanoscale physics (n.d.) http://nanophys.seas.upenn.edu/nanotube4.html. Accessed 15 Aug 2015
20. Wang Z, Ci L, Chen L, Nayak S, Ajayan PM, Koratkar N (2007) Polarity-dependent electrochemically controlled transport of water through carbon nanotube membranes. Nano Lett 7:697
21. Lu X, Chen Z (2005) Curved pi-conjugation, aromaticity, and the related chemistry of small fullerenes (C60) and single-walled carbon nanotubes. Chem Rev 105(10):3643–3696
22. Camilli L, Pisani C, Gautron E, Scarselli M, Castrucci P, D'Orazio F, Passacantando M, Moscone D, De Crescenzi M (2014) A three-dimensional carbon nanotube network for water treatment. Nanotechnology 25(6):065701
23. Talbot C (1999) Fullerene and nanotube chemistry: an update. Sch Sci Rev 81:37–48
24. Mohanraj VJ, Chen Y (2007) Nanoparticles-a review. Trop J Pharm Res 5(1):561–573
25. Okuyama K, Lenggoro W, Iwaki T (2004) Nanoparticle preparation and its application-a nanotechnology particle project in Japan. In: MEMS, NANO and smart systems; ICMENS; Proceedings; International Conference on IEEE, p 369–372
26. Trivedi R, Kompella UB (2010) Nanomicellar formulations for sustained drug delivery: strategies and underlying principles. Nanomedicine (Lond) 5(3):485–505

27. Velagaleti PR, Anglade E, Khan IJ, Gilger BC, Mitra AK (2010) Topical delivery of hydrophobic drugs using a novel mixed nanomicellar technology to treat diseases of the anterior and posterior segments of the eye. Drug Delivery Technol 10(4):42–47
28. Vadlapudi AD, Mitra AK (2013) Nanomicelles: an emerging platform for drug delivery to the eye. Ther Deliv 4(1):1
29. Quintanar-Guerrero D, Allémann E, Doelker E, Fessi H (1998) Preparation and characterization of nanocapsules from preformed polymers by a new process based on emulsification-diffusion technique. Pharm Res 15(7):1056–1062
30. Acosta E (2009) Bioavailability of nanoparticles in nutrient and nutraceutical delivery. Curr Opin Colloid Interface Sci 14(1):3–15
31. Leong TSH, Wooster TJ, Kentish SE, Ashokkumar M (2009) Minimising oil droplet size using ultrasonic emulsification. Ultrason Sonochem 16(6):721–727
32. Silva HD, Cerqueira MÂ, Vicente AA (2012) Nanoemulsions for food applications: development and characterization. Food Bioprocess Technol 5(3):854–867
33. Mohan Raj VJ, Chen Y (2006) Nanoparticles—a review. Trop J Pharm Res 5:561–573
34. Singh A, Garg G, Sharma PK (2010) Nanospheres: a novel approach for targeted drug delivery system. Int J Pharm Sci Rev Res 5(3):84–88
35. Williams L, Adams W (2007) Nanotechnology demystified. McGraw-Hill, New York
36. Drbohlavova J, Adam V, Kizek R, Hubalek J (2009) Quantum dots—characterization, preparation and usage in biological systems. Int J Mol Sci 10:656–673
37. Shirasaki Y, Supran G (2013) Emergence of colloidal quantum-dot light-emitting technologies. Nat Photonics 7:13–23
38. (n.d.) http://www.marketsandmarkets.com/PressReleases/quantum-dots.asp. Accessed 9 July 2015
39. (n.d.) http://www.prnewswire.com/news-releases/global-nanocomposites-market-is-expected-to-reach-usd-591-billion-by-2018-transparency-market-research-215820981.html. Accessed 9 July 2015
40. Pavlidou S, Papaspyrides C (2008) A review on polymer-layered silicate nanocomposites. Prog Polym Sci 33:1119–1198
41. (n.d.) http://ntechresearch.com/news/article/silver-nanowire-transparent-conductors-to-become-a-500-million-market. Accessed 9 July 2015
42. Xia Y, Yang P (2003) One dimensional nanostructures: synthesis, characterization, and applications. Adv Mater 15:353–389

Chapter 2
Construction and Building Applications

The construction industry is a particularly attractive industry for nanotechnology applications. Since it employs a diversity of building materials, nanomaterials can be employed to enhance the materials' performances, durability, longevity and sustainability. The construction and building practice is detrimental to the environment in various aspects such as electricity consumption, landfill accumulation, unhealthy aesthetics and neighborhoods. This chapter shows how nanotechnology-based building materials are playing an important role in green architectural design and construction, which has become a growing trend in many countries. Nanomaterials could be integrated with cement, concrete, or windows to conserve energy, minimize electricity bills and sanitize the surrounding atmosphere. Nevertheless, the industry is still facing many hurdles towards the wide application of nanomaterials in a cost effective manner.

Introduction

Construction and building are major contributors to the rising carbon emissions and the global climate change. In addition, the construction industry is a rapidly growing industry. The nature of the construction industry involves the application and use of a diversity of building materials. This leads to nanotechnology being rapidly adopted by the construction market in several aspects like coatings, insulation materials, and building materials (steel, cement, asphalt, glass, polymers, etc.). Nanomaterials are currently employed in cements, steel, and even windows to render buildings greener, more cost effective and safer.

Green construction has become the dominant trend in the construction industry, which means that buildings should meet certain specifications in the materials used, the processes employed, and in the behavior of the residents. These specifications aim at reducing CO_2 emissions and reducing the negative impact of the construction industry on the environment. Green architectural design and construction

M. Anis et al., *Nanovate*, DOI 10.1007/978-3-319-44863-3_2

has shifted dramatically since the evolution of nanotechnology. Many people are now seeking healthy lifestyles and environmentally friendly neighborhoods. Green building is currently experiencing a growing demand by governments, architects, and people, especially after many cities have adopted strict regulations regarding green construction. Globally buildings consume around one third of the worldwide electricity [1], with cement alone contributing to the global CO_2 emissions by around 5 % [2].

One way nanotechnology is used in construction is the concept of nanoreinforcements which involves reinforcing the body by adding dispersions that are in the nano-size scale. For example, nanoreinforcements in steel have rendered steel stronger and lighter. Nanoreinforcements in cement have rendered it more durable and cost effective. Another application of nanotechnology in the market involves making Ultra Violet (UV) absorbing, self-cleaning, and depolluting coatings for windows. Moreover, nanotechnology has been rapidly adopted in air and water purification systems. The next generation of nanotechnology applications in construction has begun to involve Building Integrated Photovoltaics (BIPV), which are solar cells that can be integrated within buildings in smart designs without affecting the buildings' aesthetics, as well as being a clean source of energy and electricity. Even electronics and sensors are being developed to be integrated with buildings. Smart curtains or windows are expected to change the way people are experiencing the way they live (imagine walls that change colors with just a click!) [1].

It is important to note that there are still a lot of barriers to the adoption of nanotechnology in building, however, there are certain drivers and enablers that if emphasized and executed will overcome such barriers. Nano-enhanced materials are more expensive than conventional materials. They yield high performances using small amounts of the materials. Thus, it is expected that with the continuous R&D (Research and Development) efforts to improve nanomaterials' performances at lower costs, the construction market will continue to increasingly adopt the nanomaterials. Proper education of the construction industry about the potential benefits associated with nanotechnology is the starting point towards the rapid adoption of nanotechnology in building materials. Although nano-based construction materials are not economically competitive compared to conventional materials, continuous efforts are undertaken to render them cost effective. In addition, strict regulations have proved to be an important surge to the adoption of nanotech in green construction. One significant example is the US LEED (Leadership in Energy and Environmental Development) which has developed strict minimum standards to simultaneously force and encourage builders to conserve energy and reduce carbon emissions.

Cases

HONG KONG SUBWAY

Hong Kong fights germs in subways.

A Japanese antimicrobial nanocoating that had long lasting efficiency offered maximum protection and elevated hygiene levels in Hong Kong subway stations.

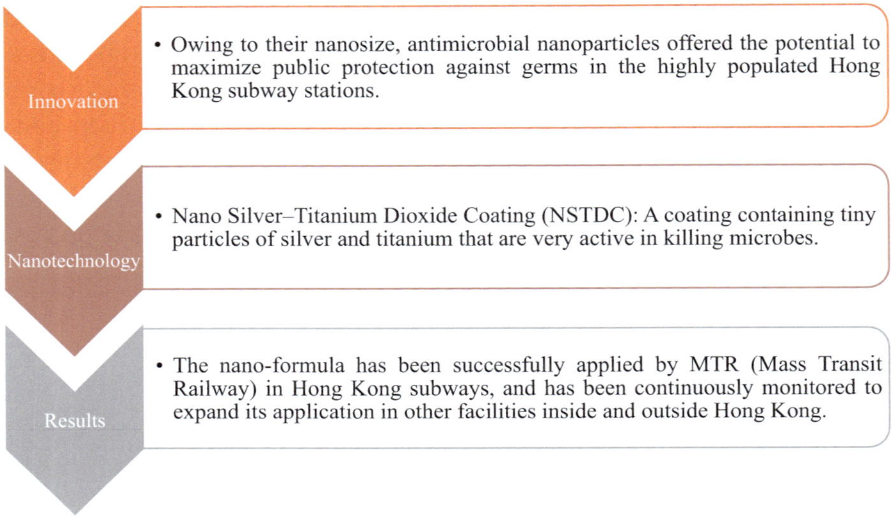

Drivers

SARS outbreak. In the spring of 2003, a large outbreak of Severe Acute Respiratory Syndrome (SARS) occurred in Hong Kong and other Asian countries. The contagion rapidly spread throughout the world. Its rapid transmission and high mortality rate made SARS a global threat for which no efficient therapy was available and proactive empirical strategies had to be developed to treat the patients and protect healthy individuals.

High risk zones. Around 2.5 Million commuters use Hong Kong railway and subway stations daily, and they are highly susceptible to dangerous microbes adhering on floors, handrails and buttons.

More cost and lower sustainability. Regular disinfection of railway stations uses harsh cleansers, which are expensive and some of which are not environmentally friendly.

Barriers

Public misconception. The US National Institute for Occupational Safety and Health (NIOSH) has classified TiO_2 as a "Potential Occupational Carcinogen", which produces a public misconception regarding its safety—workers need to wear protective clothes while spraying NSTDC, but once dried it is completely safe.

Silver resistance. There are recent investigations that are addressing issues related to microbes that can develop resistance against silver nanoparticles, which can compromise its long term effectiveness.

No international standardized testing. Lack of standardized tests to assess the long term potential environmental impact of nanoparticles (because there are different formulations of different compositions using different synthetic techniques).

Enablers

Efficiency on a broad spectrum. NSTDC is certified to be effective in killing a wide range of bacteria, viruses, and mold including the H1N1 influenza virus.

Huge investment. MTR Corporation attempted to invest 1.5 million dollars in nanotechnology products to enhance the hygiene levels of Hong Kong stations, especially that nanoparticles maximize effectiveness.

Regulated materials. NSTDC's main component, titanium dioxide (TiO_2), has been approved for use in foods by the FDA and under the Public Health and Municipal Services Ordinance in Hong Kong.

Impact

Promising results. Preliminary tests on Hong Kong subway cars coated by NSTDC showed 60% reduced bacterial infections.

Durability. The NSTDC nano formula is stable and does not require frequent replacement; therefore it is sprayed every 3 years and is checked regularly every 8 months.

Wider adoption. NSTDC, promising antimicrobial efficiency, makes it suitable to be applied in MTR-managed shopping malls, staff offices and recreational facilities to ensure the highest cleanliness standards for passengers, customers and staff.

What's Next?

MTR plans to monitor the coatings in Hong Kong in attempts to apply them in its two new UK rail franchises—London Rail and West Midlands.

HYCRETE

Palais Royale: India's first green super-tall residential building.

Hydrophobic concrete plays an important role in the rapidly growing building industry in India.

Innovation

- Concrete is porous and by time, it absorbs water causing it to corrode and crack. Hycrete admixtures (chemical mixtures added to concrete) are hydrophobic nanostructured mixtures that create a waterproofing seal on the concrete's internal capillaries preventing water penetration. Similarly, they protect steel reinforcement bars by creating protective coatings around them.

Nanotechnology

- Nanostructured cement in liquid admixtures: When the concrete hardens, the admixture turns from liquid to a solid bonded to concrete.

Results

- "Palais Royale" is the first skyscraper to achieve the platinum LEED certificate from the Indian Green Building Council.

Drivers

Rapidly emerging market. The construction boom and the damp environment in India make it an ideal customer for Hycrete. It is worth mentioning that in 2008 alone, around 30 projects made use of Hycrete.

Cost of corrosion. The US spends 276 billion dollars annually on corrosion related problems and the Federal Republic of Germany spends 4% of its gross national product on corrosion [1].

LEED motif. Builders and architects all over the world seek to earn LEED credentials. India has become an advanced entrant in the LEED race.

Environmental impact. External protective coatings on concrete are usually petroleum based and thus compromise the ability to recycle concrete. In addition, they release VOCs (Volatile Organic Compounds) into the surrounding environment.

Barriers

Still needs to be greener. Hycrete technology does not play too impactful of a role in reducing carbon emissions associated with making concrete.

Very slow adoption rate. The construction market is somewhat conservative and diffuse, especially in India: high quality construction is not yet fully appreciated on a large scale.

Risky investment. The construction market in developed countries is well established. Consequently, developing countries are the main target. However, the concepts of green nanotechnology innovations are not easily implemented or understood in most developing countries.

Uncertain feature. Some studies on Hycrete suggested that its compressive strength is reduced by 10–20 % yet is still suitable for most construction applications [3].

Enablers

India is getting rich. The construction boom in India was mainly triggered by the rapid growth in the middle-income class as well as the increasing wealth of the country.

High level testing. Hycrete went through over 10 years of independent and sponsored tests, mostly funded by US Federal Highway administration. All of these tests showed promising results. According to the *British Standard Absorption Test BSI 1881-122*, Hycrete had less than 1 % water absorption [4].

Gold certificate. Hycrete is the first material certified by Cradle-to-Cradle gold certification because it is easy to recycle and it adds LEED points to projects.

Cost and time saving. The admixture is built within the concrete that is batched at the plant not at the job site. Thus it saves costs of extra containers and materials compared to the case of applying external coatings at the job site.

Impact

Business boom. The *Palais Royale* project greatly boosted the Hycrete business. The project required 100,000 gal of Hycrete and the company gained more than one million dollars.

Cost efficiency. The installation process of Hycrete in the "Palais Royale" project was fast and cost effective because the amount of Hycrete needed per square feet was 40 % less than the amount needed for the external waterproofing membrane solutions.

Operational impact. Implementation of nanostructured admixtures in Hycrete decreases the amount dosed in concrete (1 gal per cubic yard of concrete) reducing the time, cost, and waste of operations [4].

Internal impact. The application of Hycrete is expected to enhance the longevity and durability of the skyscraper's infrastructure.

What's Next?

The underdeveloped construction market in India along with Hycrete's success in Indian projects is driving private equity firms to joint venture with Hycrete, Inc. Hycrete's achievement in India is setting it as a role model for the emerging construction markets in the Middle East.

BIONI, INC.

Bioni paints Dubai Discovery Gardens.

A 40,000 square meter residential community in Dubai "Discovery Gardens" decided to use the antimicrobial Bioni paints.

Innovation

- Bioni using nanotechnology, managed to stabilize and integrate silver nanoparticles in paints. It used the smallest possible microbicidal amount to prevent yellowish discoloration of paint.

Nanotechnology

- Silver nanoparticles bound to paint: tiny particles of silver that are very active in killing microbes.

Results

- The residential community is built with an antimicrobial durable nanocoating that is resistant to cleansing chemicals and is nonflammable [1].

Drivers

Greener long lasting alternative. Traditional antimicrobial coatings containing biocides have short life spans and slowly diffuse into the environment posing human and environmental risks.

Metropolitan growth. In Dubai, there is a continuously growing construction activity with buildings of truly immense proportions, which makes it a valuable customer for Bioni antimicrobial paints.

Desert kingdom. Although Dubai employs high quality construction work, the hot climate and high humidity create an ideal medium for the growth and propagation of mold and microbes inside buildings.

SBS. The SBS (Sick Building Syndrome) is strongly related to contaminants and pollutants released from indoor sources [1].

Barriers

Aesthetics and efficiency. The difficulty of maintaining a high enough concentration of nanosilver to efficiently exert an antimicrobial effect yet low enough to prevent yellowish discoloration.

High cost. Although Bioni antimicrobial paints in hospitals are 25 % more cost effective because they decreased the frequent use of biocides, they are more expensive in residential buildings.

Superbugs. Even though tests have shown that Bioni antimicrobial paints killed more than 99 % of Staphylococcus aureus on the paint surface, still there were no appropriate field tests carried out to test activity against superbugs (microbes resistant to multiple antibiotics).

Enablers

Successful research collaboration. Researchers at the Fraunhofer Institute for Manufacturing Engineering and Applied Materials Research in Bremen) and in Bioni CS have spent more than 10 years developing processes for manufacturing the antibacterial nanosilver particles and for incorporating them successfully into paint solutions.

Broad microbicidal effect. Bioni antimicrobial paints are volatile organic compounds, free of antiallergic, antiviral and antibacterial effects.

Nanosafety. The nanotechnology innovation in the paint ensures nanosafety by trapping the nanoparticles in a polymeric matrix, thus there is no fear of release into the surroundings. Furthermore, the coating is certified by TÜV Rheinland Signet ensuring that it is emission free and does not release any substances into the environment.

Dubai municipality testing. The national building authority of Dubai—the Dubai Municipality—conducted its tests on Bioni products yielding promising results.

Impact

Maintain building integrity. The innovative paints proved to be the best fit with building materials because they are resistant to abrasion (wet abrasion resistance class 2) and disinfection chemicals. Additionally, they are not flammable (building material class A2).

Extra cost efficiency. According to Dubai municipality testing, Bioni's nanosilver technology has low thermal conductivity and can reflect up to 93 % of the incident light, thus lowering air conditioning bills.

Growing market in Dubai. On July 13th 2009, Bioni was officially recorded in the Emirate's commercial register marking an excellent opportunity for Bioni to expand its activities in the Gulf region.

What's Next?

Although Bioni antimicrobial paints were designed to target medical facilities, Bioni they can find applications in other sensitive buildings such as children's bedrooms, schools, kindergartens, bathrooms, showers and toilets, the food industry, and retirement homes for the elderly.

SOCIÉTÉ GÉNÉRALE

Can windows save energy?

Société Générale bank in Switzerland utilized 3 M nano-based UV absorbing window films to solve the problems of its overheated indoor working environment.

Innovation

- Hundreds of ultra-thin layers made of advanced nanoceramics that can reduce heat and ultraviolet light transfer were installed. At the same time, they preserve interior and exterior window views.

Nanotechnology

- Nano-based advanced ceramics: They filter out 25–99 % of infrared (IR) and UV radiations, while maintaining the transfer of visible radiations [5].

Results

- The new windows reduced cooling costs, contributed to the environment, and maintained their healthy indoor environments.

Drivers

Nowadays' impractical alternatives. Conventional reflective films are either metal based, which interfere with electronic signals and corrode in coastal regions changing the window color, or dye-based, which fade over time.

Sustainability. Energy costs are increasing, and there is a great need for sustainable energy efficient solutions especially from the building sector which is one of the main detrimental influences on the environment.

Promising technology. The US Department of Energy (DOE) conducted a study on the top 50 commercially available energy conservation technologies. The results show that window films were ranked as a top tier technology with the highest probability of success and the fastest Return on Investment (ROI).

Barriers

Landmark status. Société Générale bank is based in a building that is considered to be Zurich's very first high rise building. Therefore, any solution to the bank's overheated indoor environment shouldn't interfere with the historical building's external features and aesthetics.

Market coverage. UV absorbing window films are not one fit for all types of windows or buildings, thus they are not equally effective in all cases.

Skeptical customer. Although the building industry serves a huge market, it is very hesitant in adopting new and uncertain technologies.

Skeptical consumer. Casual residents may underestimate the cost efficiency associated with installing UV absorbing window films until they see real cost savings.

Enablers

NANO. 3 M window films are made of nanoscaled ceramics. At this scale, materials are invisible, therefore, the films keep windows clear and less reflective compared to glass.

First patency. 3 M is the first company to earn the world's first patency in window films over four decades ago. Since then, the company has been producing reliable solutions to lower customers' energy costs and to provide an accelerated return on investment.

Marketing strategy. 3 M supports its window films with a 15 year commercial warranty, encouraging its rapid commercialization and adoption.

LEED. 3 M window films are components of LEED sustainable design where buildings can earn 2–11 points [6].

Impact

Indoor environment. Up to 99 % of total UV energy and more than 80 % of the heat producing IR energy were rejected. The bank had better and sustainable indoor temperature control. Furthermore, fabrics and furnishings were protected against UV induced damage.

Cost effectiveness. Société Générale bank enjoyed a unique and green approach to reduce energy consumption with a valuable return on investment.

Mission accomplished. The bank managed to improve the indoor working environments without compromising the specific historic and aesthetic features of the old building.

Green impact. The building industry is a top contributor to worldwide carbon emissions. Building insulations could save around 42 % of energy consumption and consequently reduce carbon emissions [1].

What's Next?

The growing regulations on the building sector to save energy drive the growth for solar control window films, which is forecast to be an 863 million dollars market by 2018. First technology providers like 3 M are likely to be market leaders, and early adopters in the construction market will enjoy a unique position [7].

PILKINGTON ACTIVE GLASS

Cleaning using forces of nature.

An innovative glass makes use of rain and sun to self-clean and resist dirt.

Innovation

- Pilkington developed nano-films made of a photocatalyst that can interact with ultraviolet (UV) rays to break down organic dirt. The innovative film also contains a hydrophilic (water loving) component that can attract water or rain on its surface forming sheets that dry without spotting.

Nanotechnology

- The technology used nano films (15–20 nm) made of microcrystalline titanium dioxide. Titanium dioxide in the presence of UV produces radicals that degrade organic pollutants to nontoxic products.

Results

- Ever since Pilkington introduced their self-cleaning glass windows, they have been experiencing accelerated adoptions of their windows and a growing customer base.

Drivers

Environment friendliness. Pilkington self-cleaning windows eliminate the need for noxious chemicals that are washed off into the environment after the cleaning process.

Consumer convenience. Self-cleaning windows are forecasted to be more rapidly adopted in construction compared to other types of smart windows: consumers can rapidly see the cost efficiency associated with the decreased frequency of cleaning windows compared to that associated with energy savings in other types of smart windows.

Emerging market. The market for self-cleaning windows is not yet mature, with few players, thus, there is a great opportunity for growth and expansion for early entrants.

Barriers

Regional limitation. The photocatalytic material used will not function properly in the presence of inorganic dirt, long spells, coastal areas and internal windows because it needs daylight.

Lack of standards. Since self-cleaning technologies in windows are new features and no quality standards are there to assess their efficiency, a consumer's decision will be reliant only on the manufacturer's claims.

High cost. Pilkington self-cleaning windows are 15–20% more expensive than conventional glass, which can negatively influence the buyer's purchasing decision.

Market upset. Glass cleaning companies are an important component of the glass windows "value chain". Thus, the introduction of self-cleaning windows will affect their businesses negatively.

Enablers

R&D. Pilkington's expertise through 5–7 years of extensive R&D on thin film technologies enabled it to successfully transform them from laboratory samples to the market.

Safe material. Titanium dioxide is publicly and governmentally accepted because it has been widely used in food related materials, cosmetics, and toothpastes.

Impact

Attraction of unique customers. Saint Pancras Station in the United Kingdom is one of the customers that has applied the self-cleaning glass. It installed the glass on its roof and building structure to reduce maintenance costs.

Technology appreciation. Pilkington was one of the four finalists for The Royal Academy of Engineering's MacRobert prize that rewards technological and engineering innovation.

New added benefit. Pilkington photocatalytic effect could naturally also depollute the surrounding air by decomposing some pollutants such as formaldehyde and ground level ozone.

What's Next?

Pilkington is working with other glass manufacturers, universities, and standards' institutes to develop standards that assess and measure the self-cleaning properties of nanocoatings in an attempt to increase the public trust (so as not to be only reliant on the manufacturer claims) and the commercial adoption of these products. Moreover, self-cleaning glass can be integrated with solar panels to reduce their frequent cleaning costs, and to make use of the same solar radiations.

LUMOTONE, INC. NANOSHUTTERS

Invisible shutter windows.

Lumotone invented new transparent shutter films that are stuck on window made opaque by the click of a button.

Innovation

- Lumotone Inc. has designed a film that can be applied to a glass surface and can be converted to opaque color upon switching it off. It can also be tailored to be colored or be used to show a projection, switched on and off using a mobile device.

Nanotechnology

- The innovation is based on a film containing polymer dispersed liquid crystals (PDLC) i.e. small nanoparticles of liquid crystals dispersed in a polymer as shown in Fig. 2.1.
- How such liquid crystals work: In a typical PDLC, there are many crystals with different orientations and configurations that scatter light giving an opaque appearance. These crystals become aligned when an electric field is applied allowing the light to pass through and giving a transparent appearance.

Results

- An efficient invisible window blind that can be turned on and off by the press of a button.

Fig. 2.1 Polymer dispersed liquid crystals used in NanoShutters

Drivers

The need for privacy. Window blinds are important house amenities to prevent people from prying into other people's lives.

New décor trends. The new futuristic décor trends constitute less furnishings, wide spaces and invisible blinds instead of curtains that can be controlled by the press of a button.

The need to save energy. By controlling the amount of light entering the room, you can ensure the room stays cool in the summer and warm in the winter thereby reducing the energy needed to maintain a room's temperature.

The need for automation. Since the advent of remote controls, computers and mobile phones people have come to rely more and more on automation. This is further increased by the modern fast pace of life. Numerous people nowadays install home automation systems to make their life more comfortable and easy.

UV blockade. NanoShutters block 99 % of UV light entering through the glass. This in turn is healthier as well as crucial to preventing the fading of colors of the interior décor and fabrics of furnishings inside the house.

Barriers

Installation. The difficulty of installation of the NanoShutter will definitely affect its success. However, Lumotone, Inc. has made the NanoShutters easy to install—as easy as putting up a poster on the wall. They also provide certified professionals who can install them for their customers.

Automation. How to make these smart windows automated and be controlled by your mobile device remotely when you are away from home.

Durability. The durability of the NanoShutter film and whether it can withstand cleaning by household solvents is a limitation. This explains why such films are protected by an upper layer to prevent them from being scratched or destroyed. In addition, the company provides directions as how to clean such smart windows; which solvents are allowed or not allowed.

Enablers

Research. The founders of the company were originally researchers and engineers in the field of nanotechnology in the University of Waterloo, Canada, giving them credible backgrounds to design such products.

NanoShutters that can be integrated with home automation systems. Lumotone developed their nanoshutters to be compatible with any home automation system.

Mobile applications for the NanoShutters. Lumotone has also designed mobile applications and computer software that enable the users to control their shutters in or away from their homes.

Auto-scheduled NanoShutters. Lumotone has made its shutters to be schedulable by computer or mobile application so a user can sleep in shaded windows and wake up in bright sunlight early in the morning to go to work.

Impact

The efficiency of NanoShutters made them invade other market sectors including automotive and aircrafts. These smart windows help to control the inside cabin temperature as well as prevent screen glare when watching videos.

What's Next?

Lumotone is further developing its technology to produce pixels and transparent electronic displays.

References

1. Elvin G (2007) Nanotechnology for green building. Green technology forum
2. (n.d.) http://www.engineeringcivil.com/nanotechnology-in-civil-engineering.html. Accessed 15 July 2015
3. http://www.hycrete.com/about-hycrete/overview/. Accessed 15 July 2014
4. Rosenberg D, Veramallay S, Horan T (2008) Environmentally intelligent integral waterproofing and corrosion protection for concrete. Clean Technology. p 182–184
5. Interwest Distribution Company (n.d.) https://interwestdc.wordpress.com/tag/window-tint/. Accessed 18 July 2014
6. (n.d.) http://manhattanwindowfilm.com/. Accessed 18 July 2014
7. (2014) Opportunities and challenges for solar control films. *LuxResearch*. https://portal.luxresearchinc.com/research/report_excerpt/17183. Accessed 18 July 2014

Chapter 3
Manufacturing Applications

Manufacturing has proven to be one of the most important sectors in an economy because it allows the country to be self-sufficient. Through manufacturing, a country can rely less on imports and can earn foreign currencies from exporting its manufactured products. Because it is such a vital sector for a country's development, intensive research and money have been invested in this sector to address its limitations as well as devise ways to help the manufacturing process be more productive and time-efficient. Nanotechnology proved to have great potential in the arena of monitoring and enhancing manufacturing processes. The role that can be played by nanotechnology is comprehensively discussed in this section.

Introduction

Even though manufacturing supports a country's economy, national infrastructure, commercial innovation and defense, there are a lot of problems with manufacturing such as real time monitoring of processes to avoid unwanted results, quality assurance and control of the products being formed, efficiency, power and resource consumption and hazardous wastes and emissions. With the advent of new technologies, more research is being done on their implementation to address the problems encountered in manufacturing. A new era of manufacturing that will most likely appear in the next few years is 3D printing and the technology of additive manufacturing that is thought to reduce the environmental impacts of conventional manufacturing. It has definitely proved itself in the manufacture of implants, prosthetic joints and dental components [1]. Other commercialized areas where nanotechnology proved itself in manufacturing include the use of sensors for real time monitoring of the particular manufacturing process at hand, coatings for prevention of bacterial growth or for insulation and anti-wear or lubrication purposes for machinery parts. Sensors during the manufacturing process include sensors to track temperature, pH, humidity, microbes, toxins, contaminants and mechanical strains or possible breakdowns [2, 3].

M. Anis et al., *Nanovate*, DOI 10.1007/978-3-319-44863-3_3

Cases

INDUSTRIAL NANOTECH, INC.

Transparent insulating coatings.

A new nanocomposite coating that has very low thermal conductivity has been developed to be used for coating equipment and machines.

Innovation

- A new industrial low-thermal conductivity coating for machines, equipment and their parts that is capable of maintaining its function at high temperatures, impacts and other drastic industrial conditions.

Nanotechnology

- A nanocomposite called Hydro-NM-Oxide that is made from silica nanogels dispersed in a film forming polymeric resin matrix.

Results

- This innovation has enabled the making of clear, nearly transparent, high-performance coatings that can be easily applied by spraying to coat steam boilers, heat exchangers, industrial machines, piping, tanks, ovens, dryers, walls, ceilings and sky lightings. Such coatings were not only found to have very good heat insulation properties but also excellent corrosion resistance and resistance to molds and moisture. The implementation of such a high-performance product was further widened to include applications in construction, housing, nuclear plants and offshore oil drills.

Drivers

A lot of heat is lost or gained during heating and cooling processes in manufacturing. The manufacturing processes' efficiencies usually depend on conservation of energy. In order to avoid loss of energy in the process this requires insulation.

Contamination of final product with machine's corroded parts. The moving machine parts in the food, beverage and pharmaceutical industries, e.g. during grinding of food, can lead to erosion of the machine part into the food or drug which would lead to contamination and toxicity. This nanocoating, however, acts as a seal over the machine parts preventing corrosion and leakage into the products.

Molding may occur during processing of foods or biomaterials leading to their contamination. The processing of foods or other materials that are prone to rotting can lead to deposits in the machines and piping that, after sometime, will attract bacteria and fungi to grow. Such bacteria and fungi can leak into the products as they are being produced. The use of a coating that has antibacterial and anti-molding properties will prevent such an event from happening.

Barriers

Safety of nanomaterials if leached into the final products. There is the problem, however, of designing coatings that don't leach nanomaterials into products as there is an ongoing debate regarding the safety of nanomaterials to human health [4]. Industrial Nanotech, Inc. though, has proven that its coatings withstand high temperatures and impacts as found in the test results on its website.

Application of the nanocoating. Whether the application of the nanocoating requires breaking apart the machine parts and then reassembling them or whether it is just spraying application is considered a limitation. In Nanosulate®, the coating can be applied by brush, roller or paint sprayer.

Enablers

Experience. Industrial Nanotech, Inc. has a history and valid experience in making protective nanocoatings. It has been conducting research, developing and selling coatings since 2004.

Awards. Due to the excellence of its product, it has received several awards including the Buildings Magazine Award, Builder News Award, Building Products Award, Qualified Remodeler Award and Journal of Architectural Coatings Award.

Testing data. Industrial Nanotech, Inc. has carried out numerous tests and published results on its patented coatings including thermal insulation data, mold resistance data, corrosion prevention data, corrosion under insulation, lead encapsulation data and environmental data.

Cost efficient products and customization. The advantage of Industrial Nanotech, Inc. is its ability to customize its products to meet different needs in different fields. It targeted, not only industrial, but also construction areas and provided different packs of the products to meet different applications and methods of implementation.

Impact

Due to the cost effectiveness of the products and their efficiency, the company was capable of holding a large market share across the globe. The products have been utilized in the textile, oil and gas, food and beverage, building and construction industries and many more, winning itself excellent customer testimonials for its products.

What's Next?

The company can add more features or properties to its coatings to target new customers like those in the sound insulation or solar energy fields.

ALPHABET ENERGY™

Thermoelectric power generation.

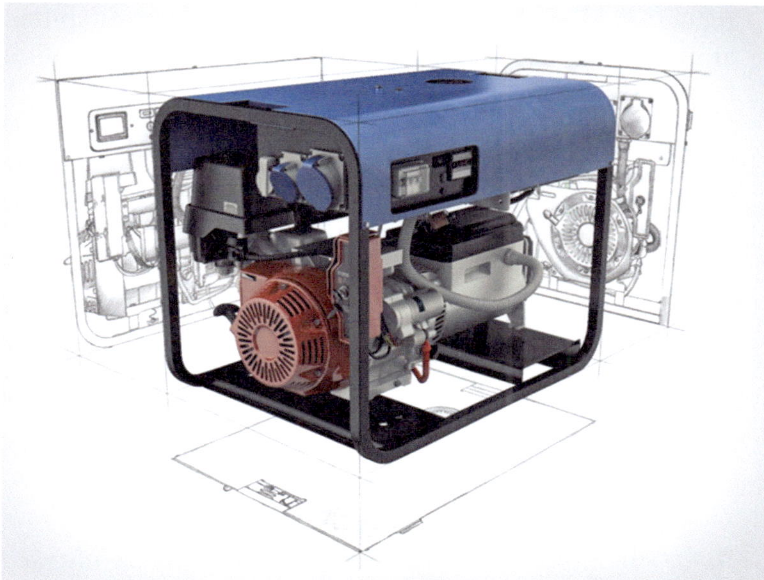

Alphabet energy was successfully capable of designing new thermoelectric generators that are capable of generating an electric current when exposed to a temperature gradient.

Innovation

- The innovation is a new energy recovery method from waste: hot exhausted gases produced in the oil industry, in manufacturing processes or by cars. The heat in the hot gases is used to generate electricity with the aid of thermoelectric materials.

Nanotechnology

- Thermoelectric materials are materials whose electrons move along a temperature gradient creating an electric current. Thermoelectric materials are materials with high electrical conductivity and low thermal conductivity eg. Lead Telluride (PbTe), Bismuth Telluride (Bi_2Te_3), and Silicon Germanium (SiGe). Such materials, however, are quite expensive. A way to improve the thermoelectric behavior of cheaper semiconductor materials is by lowering their thermal conductivity achieved through nanotechnology. The nanoscale features of dispersing nanoparticles, nanowells, nanowires or creating nanointerfaces help to reduce thermal conductivity without affecting electrical conductivity.

Results

- Alphabet Energy™ was able to create a thermoelectric-based energy recovery technology that is cheap, light in weight, efficient, reliable, can recover energy from waste gases and requires little maintenance as there are no engines involved.

Drivers

Reduction in operation costs. All industries want to achieve a reduction in operation costs and increases in profitability. The ability to recover any lost energy and reuse it in operation will help them cut down on a lot of their costs.

Sustainability. The 1970s energy crisis was an alarm to the world about its dependability on nonrenewable energy sources and how those will eventually be depleted.

Low maintenance energy recovery solutions. Since there are no moving parts or motors involved, Alphabet Energy's thermoelectric generators require very little maintenance unlike the conventional energy recovery systems.

Barriers

Achieving maximum electrical conductivity while keeping a minimum thermal conductivity in the same material. This barrier was overcome through the nanoscale design of the nanofeatures inside the material to reduce its thermal conductivity without affecting its electrical conductivity.

A thermoelectric circuit that can withstand several thermal cycles. The high temperatures of the exhaust gases can affect the electric circuit in which the thermoelectric material is placed which requires special design in order to withstand such high mechanical stresses.

Enablers

Patents. Alphabet Energy has over 60 patents and many more publications. Its cofounders are scientists experienced in the field of thermoelectric power generation.

Funding and partnerships. Alphabet Energy has been funded with over 30 million dollars from investors such as TPG Biotech®, Claremont Creek Ventures, California Clean Energy Fund and Encana Natural Gas. Moreover, the US Dept. of

Energy, the US Army and the US Air Force have contributed 2 million dollars.

Experience. Alphabet Energy has been working in the energy recovery thermoelectric based generators since 2009 and has been accumulating experience and resources ever since. It has also developed a 7900 square-foot manufacturing facility in California.

Awards and publicity. The company has received several awards for its high performance products and has been mentioned in several news anchors and several agencies such as Bloomberg, Forbes, National Geographic and Fortune.

Impact

The company's outstanding research in nanotechnology and product development has enabled it to earn many awards as well as embark on major projects in partnership with big oil companies.

What's Next?

Alphabet Energy will start to commercialize their product globally rather than solely the United States.

DYNALENE

Heat transfer nanofluids.

Dynalene used colloidal suspensions of nanoparticles to improve the heat transfer efficiency.

Innovation

- A new low conductivity heat transfer fluid used to cool or heat processes and electronics in manufacturing.

Nanotechnology

- The use of colloidal suspensions of nanoparticles like nanodiamonds to improve the heat transfer efficiency of the fluids [5]. Such nanofluids have higher heat conductivity properties and heat transfer coefficients which depend on the materials used and the rheological properties of the fluid [6]. The higher thermal transfer ability is due to their larger interacting surface areas. An added advantage of nanofluids is that their heat transfer properties can be made controllable and responsive to stimuli [7].

Results

- Dynalene was able to make a high efficiency transfer fluid with high heat transfer coefficient that can be used in cooling and heating processes. The Dynalene fluid has also added nanoparticles to prevent corrosion of the surfaces they come in contact with during operation which gives it an added advantage of extending the life of equipment and machinery used. The low electrical conductivity designed in the nanofluid has made it suitable for the cooling of electronics and fuel cells. Moreover, since the Dynalene fluid is water based, it was proven to be nontoxic and nonflammable.

Drivers

Higher efficiency heat transfer fluids. Due to the larger interacting surface areas of nanoparticles, they are able to transfer more heat.

Stable suspension fluids. Suspensions of nanoparticles are more stable than the use of larger particle slurries such as those used in conventional heat transfer fluids

Reduced pumping power. As nanoparticles transfer more heat, lower circulation rate and lower pumping power are required compared to conventional heat transfer fluids.

Reduced clogging. With conventional heat transfer suspensions clogging can occur. Comparatively, with nanoparticles, no clogging occurs.

Adjustable properties. By modifying the type of nanoparticles used, their shape, their size and the rheological properties of the fluids, the heat transfer efficiency can be manipulated as well as other properties like electrical conductivity.

Barriers

Selection of the nanoparticles. Extensive research is required to decide upon the best nanoparticles and nanofluid formulations for heat transfer.

Instability of the nanofluids. Nanofluids are always subject to coagulation of their particles due to the high surface energy of the nanoparticles. This therefore necessitates an optimal surfactant addition to prevent this while maintaining the heat transfer efficiency.

Enablers

Experience. Dynalene has been working in the field of developing and enhancing heat transfer fluids since 1993.

Grants and awards. The company has received over four million dollars in research grants to develop their products. These include the US Department of Energy Grant for Fuel Cell Coolants and the US Department of Defense Grant for Hybrid Nanoparticle Utilization in Cooling Systems.

Customization. Dynalene customizes products according to each customer's requirements while meeting the applications the product is designed for.

Wide product profile. The company has over 40 patented products with varying properties from ultra-low-temperature to ultra-high-temperature heat transfer fluids.

Laboratory services. In addition to the maintenance and customer inquiry services it provides, Dynalene also provides laboratory services and resources to its customers including quality control, stability studies, material compatibility and corrosion studies. This is enabled by the extensive research facility owned and developed by the company over the years comprising four laboratories and many expensive testing and characterization equipment.

Publications, conferences and university affiliations. It has a multitude of publications in respectful journals and conferences as well as affiliations with universities and researchers including Lehigh University, Pennsylvania.

Impact

Due to frequent research and the continuously using new materials and technologies in addition to the outstanding performance of its products, the company grew rapidly, ranked by Inc. magazine as one of the fastest growing companies. It was also nominated for the 2014 Governor Impact Award.

What's Next?

The company has potential to expand in many areas. One opportunity is to make its nanofluids controllable using magnetic fields and magnetic nanoparticles. This would develop the application profile of the fluid.

AEROTECH®

Nanopositioners.

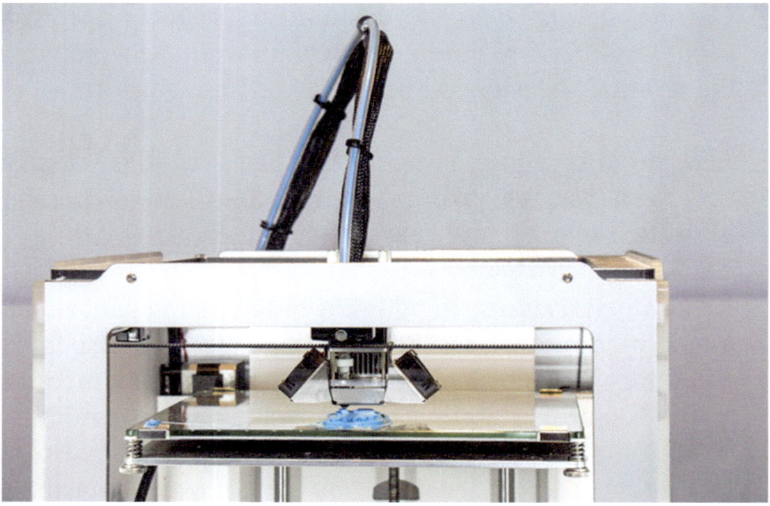

Aerotech designed piezoelectric positioners that are capable of controlling motion down to the nanosize scale.

Innovation

- Aerotech developed an advanced tool for extremely precise controlling of motion and positions down to the nanoscale level.

Nanotechnology

- In order to achieve control of motion at the nanoscale level, Aerotech used piezoelectric crystals of high purity and small grain size as well as the use of highly accurate sensors and electronic circuits that minimize noise. It also used its piezo actuators in a closed loop mode to stop hysteresis and nonlinearity.

Results

- The high quality design of the nanopositioner has enabled the company to produce products that are precise and repeatable with extremely good resolutions (to the nano level) which has enabled Aerotech to competitively enter the market and serve a wide range of fields including manufacturing (additive manufacturing), medical, research, aerospace and several others.

Drivers

Precise control of movement especially in electronics and additive manufacturing. The two major fields where precision is necessary is in the electronic circuits and additive manufacturing fields, where the specific positioning and attachment of components can affect the functionality of the finished product [8].

Miniaturization of electronic devices. Most devices nowadays have become smaller in size and will continue to shrink. Such miniaturization of devices requires minute circuits that are designed with high accuracy. This can only be achieved by manufacturing at the nano level.

Barriers

High complexity of the technology. This meticulousness, reproducibility and resolution of motion require an intricate design of circuits and components. Since Aerotech has been working in the field for over 40 years, it has access to experienced engineers and researchers as well as credible resources and facilities.

Enablers

Experience. Aerotech has been conducting research and developing in the field since the 1970s (45 years), thus it has been gaining mechanical and electrical experience throughout the years as well as establishing a wide customer base.

Manufacturing facilities. The company has advanced manufacturing facilities including clean rooms and vacuum compatible platforms.

Awards. Aerotech has received more than 20 awards for its excellence and outstanding products which has helped it sustain its reputation in the market.

Certifications. Due to Aerotech's large number of patented technologies and high quality products, it has received an ISO certification for quality standard in 1995.

Collaborations. In order to maximize benefits from its collaborations, Aerotech designed an integration program where the collaborating companies would put in a good word for Aerotech in their networks in exchange for training and discounts.

Worldwide service and support. Aerotech has been able to prove its presence around the globe. It is present in china, Taiwan, Japan, UK, USA, Germany, Canada and Mexico. It's also been able to provide a 24/7 support to its customers worldwide.

Customization. Aerotech has the technology to customize its controllers to fit to a customer's desired application.

Impact

Due to its high quality products, Aerotech has managed to invade the globe. Its high precision and nanoscale resolution as well as its refined technical support has enabled it to be one of the best in the market.

What's Next?

Aerotech's vast experience and facilities can enable it to advance its technology further to reach atomic scale.

FIBERIO®

Nanofiber manufacture.

FibeRio® invented a new technology for making nanofibers based on the centrifugal force principle.

Innovation

- Designing a new manufacturing tool called Forcespinning® machine that enables the low-cost high-production rate of nanofibers to be applied in textiles, solar, filtration and medical industries.

Nanotechnology

- FibeRio has developed a new technology for making nanofibers on a large scale, at low cost, with high productivity and without the disadvantages of methods such as electrospinning and meltblown. This technology is based on the centrifugal force principle where the polymer is drawn into nanofibers with a high speed rotating spindle.

Results

- The use of Forcespinning® allows large scale industrial production of nanofibers at low cost. It also uses less solvents compared to the other methods. It also has the advantage of reducing energy consumption because it does not use hot air jets.

Drivers

Increasing applications of nanofibers. The applications of nanofibers are becoming more well apprehended these days with applications in textiles, packaging, drug delivery, tissue engineering and energy applications. This necessitates the presence of a manufacturing method that is cheap and can be scaled up to meet industrial needs.

Low cost large scale production. Forcespinning can be used for large scale production as it does not involve the high electric field of electrospinning which would be dangerous on large scales. Additionally, it does not require the high speed hot jets of air in melt spinning which could also be dangerous and non-economical on large scales [9].

Solvent usage and recovery. Forcespinning requires less solvent than electrospinning and can be used for spinning solids which is an advantage over electrospinning and eliminates the solvent recovery step or the presence of residues of solvent in the product.

Low dielectric constant materials spinning. Materials with low conductivity are difficult to spin into nanofibers with the electrospinning process which puts numerous restrictions on the use of such materials. This, however, is not evident with Forcespinning which uses centrifugal rather an electric force to draw nanofibers [10].

Barriers

Design. The design of the device that produces the correct centrifugal force to draw the fibers requires advanced engineering expertise in order to make high-efficiency products. In FibeRio, this was not a problem as it has experienced personnel ever since its establishment in 2009.

New factors affecting fibers. This new technology to make nanofibers using a different type of force to spin them has not been studied as much as the previous technologies used: electricity or jets of hot air. However, there are increasing numbers of papers being published now about the possible factors that can affect the properties of the fibers using Forcespinning.

Enablers

Effectiveness. FibeRio's products proved to be more effective than other conventional methods used as published by several papers [9, 11]. They were also proven to be more sustainable due to the fact that they use very small amounts of solvents and less energy than other methods.

Publishing and conferences. FibeRio published countless papers since it was founded and has attained several patents. The company also presented its work about the importance of nanofibers at myriad conferences.

Collaborations. In 2015, the company made a collaboration with a big textile company to use its machines. It also collaborated with Saudi Arabia Basic Industries Corporation and Aster Capital for investments.

All scales. The company provides Forcespinning machines for research and development as well as for large scale industrial production to meet all consumer demands and widen their customer base.

Impact

FibeRio, with its high efficiency products, has been highly regarded among manufacturers. Moreover, it has received many credible awards, one of which is its being named in the prestigious 2013 Global Cleantech 100 (a global market intelligence and consulting firm).

What's Next?

It would be beneficial for the company to incorporate other technologies like applications of electric potential with Forcespinning to see the effect and see if the efficiency is further increased.

TRIBOTEX

Self-assembling lubricant.

TriboTEX invented a self-assembling nanocomposite lubricant that heals corroded parts.

Innovation

- A new lubrication system that's for industrial machines' gears, ball bearings and other moving parts.

Nanotechnology

- A self-assembling nanostructured oxide-carbide ceramic coating that is formed of two layer components: a bottom layer that sticks to the moving steel gears or bearings and a top slick lubricating layer that reduces friction.

Results

- A high efficiency coating that is capable of protecting against friction and wear. The coating's simple application has allowed a reduction of maintenance costs as well as the avoidance of the troublesome machine disassembling and remote maintenance. The coating also allows for failure intervention by detecting initial micro pitting or an increase in running temperature before catastrophic failure.

Drivers

Machine wear and corrosion during operation. Due to the friction between machine parts moving against each other, such high shear stresses can cause corrosion and pitting of the surface.

High maintenance costs for corrosion. The cost of replacing worn out machine parts, sometimes the entire machine, is usually excessively high. For example, gearboxes of turbines are designed and financed to last for 20 years, but their average lifespan is only 7 years. Replacing a turbine gearbox costs $100,000 to $250,000, while routine maintenance and repair costs amount to an additional $10,000 annually.

Leakage of corroded surfaces into products. This is especially important in the case of machines that are used for grinding of food or the likes where any worn out surface-bits that fall into the food will reach our bodies causing toxicity or diseases.

Dangerous heating up of machine parts due to friction. Friction creates a lot of heat and this increase in temperature of the equipment may start a fire.

Barriers

Technology. Fabricating such coatings requires a lot of experience and research to achieve. This, however, was overcome by the company's mixture of scientists and engineers working together.

Safety of nanomaterials. There have been published reports about the leakage of nanomaterials from coatings and about that their potential toxicity. Therefore, measures have to be taken to avoid such coatings from being scraped off the surface.

Compatibility with different surfaces. Not all coatings comply with all surfaces. Coatings have to be designed to stick to the surface which would vary if the surface is metallic or non-metallic. In the case of DuoLife, it is only compatible with metallic surfaces.

Efficiency in harsh working conditions. Coatings can lose their efficiency if subjected to high temperatures or pressures. For DuoLife, its nanoparticles are fabricated to be activated by high pressure and temperature.

Enablers

Performance improvement. DuoLife reduces friction by ten times and increases the lifetime of machines by 20% which makes it a very valuable product in reducing costs of replacing machines or their parts.

Ease of application. DuoLife can be applied without breaking apart the machine parts. The machines are set to operate to reach their working temperature and then are turned off. The lubricant system is applied and the machine turned on again for an hour.

Repair not only protect. DuoLife doesn't only lubricate and protect against wear, but it also serves to repair worn machinery as the coat self-assembles on the surface to fill in any wear grooves.

Impact

TriboTEX with its products has been able to obtain funds from the National Science Foundation and other institutions although it was just a small startup company. The efficiency of its products and the number of patents it has acquired TriboTEX to stand its ground in the market.

What's Next?

The next step for TriboTEX is to produce its products in larger volumes and invade the worldwide market with distribution agencies around the world. It can also make different tailored coatings to coat different types of surfaces.

References

1. Kreiger M, Pearce JM (2013) Environmental life cycle analysis of distributed 3-D printing and conventional manufacturing of polymer products. ACS Sustainable Chem Eng 1(12):1511–1519
2. Fang B, Jiang Y, Rotello VM, Nüsslein K, Santore MM (2014) Easy come easy go: surfaces containing immobilized nanoparticles or isolated polycation chains facilitate removal of captured *Staphylococcus aureus* by retarding bacterial bond maturation. ACS Nano 8:1180–1190
3. Arora K, Chand S, Malhotra BD (2006) Recent developments in bio-molecular electronics techniques for food pathogens. Anal Chim Acta 568:259–274
4. Johnson D (2010) Determining nanotechnology health effects. ISHN 44(1):27
5. Kakaç S, Pramuanjaroenkij A (2009) Review of convective heat transfer enhancement with nanofluids. Int J Heat Mass Transfer 52:3187–3196
6. Chen H, Witharana S, Jin YS, Kim C, Ding Y (2009) Predicting thermal conductivity of liquid suspensions of nanoparticles (nanofluids) based on Rheology. Particuology 7:151–157
7. Shima PD, Philip J (2011) Tuning of thermal conductivity and rheology of nanofluids using an external stimulus. J Phys Chem C 115:20097–20104
8. Zhai Y, Lados DA, LaGoy JL (2014) Additive manufacturing: making imagination the major limitation. JOM 66(5):808–816
9. Sarkar K, Gomez C, Zambrano S, Ramirez M, de Hoyos E, Vasquez H, Lozano K (2010) Electrospinning to forcespinning. Mater Today 13(11):12–14
10. Subbiah T, Bhat GS, Tock RW, Parameswaran S, Ramkumar SS (2005) Electrospinning of nanofibers. J Appl Polym Sci 96(2):557–569
11. Padron S, Fuentes A, Caruntu D, Lozano K (2013) Experimental study of nanofiber production through forcespinning. J Appl Phys 113(2):024318

Chapter 4
Clean Energy Applications

Most energy sources that the world relies on are non-renewable sources that will be depleted in the very near future. Our energy consumption is rising at an exceptionally fast pace due to the increase in the world population coupled with the widespread urbanization noticed all over the globe. This necessitates new alternative methods of generating or conserving energy. How nanotechnology plays an important role in such a process will be discussed more in this section.

Introduction

As the world population continues to grow, energy consumption also increases. This has lead governments to seek different ways to either reduce consumption of energy or increase production of sources of energy especially renewable ones. Nanotechnology has been extensively researched for ways to be harnessed for such purposes and was found to be an efficient tool in serving its purpose. Nanotechnology has served to reduce energy consumption in a number of ways including making better insulation layers and materials as well as enabling the production of strong lightweight materials that can be used for more efficient and less energy consuming transport. Nanotechnology has also directly affected consumption by improving energy conversion seen in more efficient combustions and electric light bulbs. Quantum dots were used and manipulated to achieve the highest energy conversion from electricity to light as well as to produce the best colors, making them of use in the new display screens and light bulbs [1, 2]. Metal nanoparticles also help to improve the efficiency of the burning of fuels by increasing their energy density, shortening the ignition delay, reducing the burning time and acting as a very high surface area catalyst that maximizes the contact between the fuel and oxidizing agent. It also improves the fuel's injection velocity into the

M. Anis et al., *Nanovate*, DOI 10.1007/978-3-319-44863-3_4

combustion chamber [3]. Nanotechnology has also allowed increased energy production. An example is the improved efficiency of fuel extraction from oil drilling sites through the use of nanofluids to squeeze the oil out and nanobots (sensors) to investigate the wells and their performance [4]. Nanotechnology has also allowed the capture of solar energy in solar cells and improved the efficiency of its conversion to electricity [5]. It has also improved energy storage in batteries and capacitors [6, 7].

Cases

HYDREXIA-SOLID STATE HYDROGEN STORAGE

Smaller lower pressure hydrogen tanks.

Hydrexia designed new hydrogen storage systems that are smaller in size, transportable, cheaper, safer and with a much higher storage capacity.

Innovation

- Hydrexia was able to develop low pressure hydrogen storage systems with high storage capabilities. These systems were designed to be safer and cheaper than the conventional compressed gas systems.

Nanotechnology

- Hydrexia designed nanostructured magnesium alloys and nanocomposites with transition metals to store hydrogen as magnesium hydride. Nanostructuring and mixing with transition metals (act as catalysts) serve to modify the sorption/desorption kinetics of hydrogen and, hence, lower the operating temperature [8].

Results

- Hydrexia was able to design a low cost (half the cost of conventional compressed hydrogen systems), hydrogen storage system that can be customized to optimize flow rate requirements of the different customers as well as their storage space requirements (up to four times less than conventional compressed storage). These storage vessels are also safer due to much lower pressure than that used in conventional compressed storage. They were also able to provide mobile storage system designs that can deliver more than three times the mobile conventional systems.

Drivers

Using hydrogen as an environmentally friendly source of energy. The use of hydrogen as an energy source has been long researched and developed due to its high energy content (142 MJ/kg, three times that of gasoline) and its environmentally friendly nature (it produces water on combustion). The need to use hydrogen as an alternative energy source has increased due to the increasing environmental pollution that the world is living in [9].

The disadvantages of high pressure gas storage. High pressure or compressed gas storage has many disadvantages such as lower energy content per unit volume (4.4 MJ/L at pressures as high as 70 MPa) and is unsafe due to the pressures used in the storage vessel as well as the energy required and consumed to compress the gas during manufacture [10].

The disadvantages of liquid hydrogen storage. Storing hydrogen as a liquid has much higher energy than the compressed gas form (8.4 MJ/L), however, it requires extensive cooling, up to −252 °C, which is an expensive process in addition to the fact that there will always be hydrogen gas lost due to inevitable evaporation.

Advantages and safety of solid state storage of hydrogen. Solid-state hydrogen storage has many advantages: it is safe due to low pressures used, more hydrogen can be stored therefore giving a much higher energy density than the two conventional methods and its sorption and desorption rate can be controlled [11].

Barriers

Composition optimization. Metal hydride storage is based on alloys of two or more metals where at least one has to form a stable metal hydride and the other an unstable metal hydride. As a result the exact composition of the different metals in the alloys will play an important role in the storage capacity, hydrogen discharge capacity, operating temperature, electrochemical properties and the lifecycle of the alloy [12].

Cost of the alloy. Most hydrogen storage alloys are expensive to fabricate and have low gravimetric hydrogen storage densities (less than the standards put by United States Department of Energy) and, thus, are not suitable as energy sources especially if used for automotive applications [13].

Poor kinetics and high operating temperatures. Although the use of Mg alloys are much better than other alloys in terms of storage capacity, they have the disadvantage of having poor thermodynamics and they would require high temperature to operate, which hinders their actual use.

Enablers

Use of mg alloys. Mg alloys have many advantages including a high storage capacity, being abundant and inexpensive. Hydrexia was able to collaborate with a Chinese manufacturer of Magnesium alloys to reduce the cost of its supply chain.

Nanostructuring. In order to improve the thermodynamic properties of the alloy and the sorption/desorption kinetics of hydrogen, thus the operating temperature, it was found out that the inclusion of different transition metals at the nanoscale in a composite or alloy form helps to improve such kinetics by acting as catalysts [14, 15]. Hydrexia has employed nanostructuring to lower the operating temperature.

Compliance with standards and testing. Hydrexia's products are designed and manufactured in compliance with regional codes, standards and certifications for pressure vessel and electrical systems. They were tested by a major industrial gas company in Australia and were approved for commercial use in 2014.

Awards. Hydrexia received the Queensland's Engineering Excellence Awards of 2012 for designing and making a storage system that can store 22 kg of hydrogen gas—a significant achievement over the conventional standard compressed cylinder which stores 600 g.

Funding and investments. Due to its excellence in the field of hydrogen storage, Hydrexia was capable of obtaining several funds and investments to continue development and commercialization including a total of 13.50 million dollars investment from Air Liquide, Southern Cross Renewable Energy Fund.

Impact

Due to the efficiency and competitiveness of its products, the company was able to prove itself in the market commercializing its product, receiving awards and raising funds.

What's Next?

Hydrexia can further develop its products and storage systems to be used for cars in an efficient way.

GASP SOLAR NANOWIRE SOLAR CELLS

Wires that trap light!

Gasp Solar has designed new solar panels made out of nanowires that have a higher light entrapment efficiency.

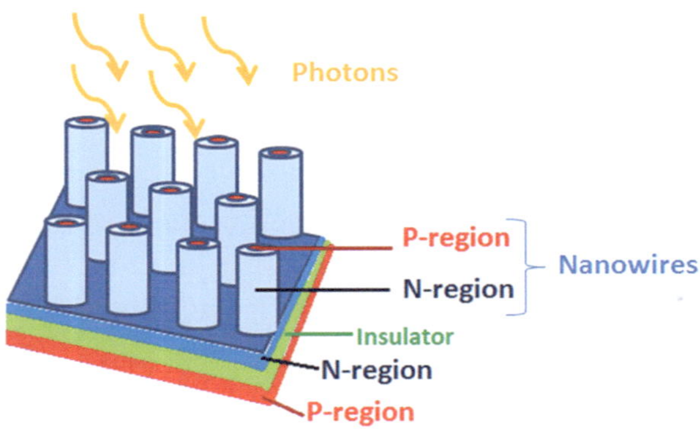

Fig. 4.1 Nanowire solar cell

Innovation

- A new solar cell that is two times more efficient and ten times less expensive than the conventional solar cell.

Nanotechnology

- It used nanowires on top of silicon to increase their light entrapment efficiencies and reduce their reflective losses [16]. Figure 4.1 shows how the nanowire solar cell look like.

Results

- Their nanowire solar cells ensure the highest possible light collection efficiency, a build-in light concentration, a near perfect black surface, and an increased acceptance angle.

Drivers

Depletion of nonrenewable fuels. The world's energy consumption is increasing by 2.3 % every year. Sources of energy, according to the International Energy Agency [17], include oil 31.4 %, coal 29.0 %, natural gas 21.3 %, biofuels and waste 10.0 %, nuclear 5.8 %, and 'other' (hydro, peat, solar, wind and geothermal power) 1.1 %. Since oil and coal are the major source of energy and are non-renewable, they can become depleted. This necessitates us to develop more renewable sources like solar, wind, hydro and geothermal.

Environmental pollution. The use of fossil fuels is one of the leading causes of environmental pollution around the globe [18]. The effects of environmental pollutants has been well comprehended in the past few years including how they affect the climate, human health as well as the biodiversity of the ecosystems. This has led to an increasing need to look for alternative clean sources of energy including wind, solar and other sources of energy.

Barriers

Cost of making a silicon solar cell and large scale implementation. It is quite expensive to make a silicon solar cell due to the extensive purification and processing of the silicon wafer.

Low efficiencies. Conventional solar cells have low efficiencies (about 20 %) and improving the efficiency is difficult especially on large scale implementations.

Enablers

Use of nanowires. One of the sources of costs for solar cells is the extensive purification process of the silicon wafer to achieve an accepted performance. Using nanowires solves this problem by orthogonalizing the light absorption and charge separation while allowing for a more efficient light entrapment [16].

Experience. Gasp Solar has been building up on research experience since 2006 when it was founded under a different name. When the company switched to Gasp Solar, it maintained its same staff, know-how and intellectual property.

Collaborations. Gasp Solar collaborated with the government through SEED Capital Denmark, the University College London, the University of Copenhagen and the Technical University of Denmark.

Impact

Due to its experience and technology, it was able to fabricate high efficiency solar cells that are ready to be commercialized on a large scale. Due to the quality of the product patented, the company is being supported by the government.

What's Next?

The next step for this company is to widely commercialize its products, especially to countries with plenty of sunlight.

GRAPHENE NANOCHEM OIL MINING NANOFLUIDS

Fluids that squeeze oil from pores!

Graphene Nanochem was able to produce nanostructured fluids that have special rheological properties that enable it to leak through well pores and push the oil out.

Innovation

- Graphene Nanochem has designed environmentally friendly non-toxic fluids that aid in the onshore and offshore drilling process. These fluids are designed to perform in a wide range of different drilling environments including high temperatures and high pressures.

Nanotechnology

- It used graphene and graphene oxide smart nanofluids whose viscosity and rheological properties are modified by the passage of an electric current.

Results

- Using Graphene smart fluids, Graphene Nanochem was able to produce excellent successful drilling fluids that help to reduce well formation damage, produce thinner filter cakes, reduce fluid loss into rock formation pores, preserve original rock formation pressure, reduce wear on drilling tools and reduce negative impact to the environment.

Drivers

Decrease in available hydrocarbon sources and increase in extraction costs. Due to the decrease in the oil and gas reservoirs, there has been a necessity to use new techniques to improve the productivity of a well. A lot of the conventional methods are costly and have low productivity. This has led to the researching of new techniques involving nanofluids, especially low cost ones to improve productivity [19].

Oil price drop. The drop in the price of oil had put demand for new ways to reduce the cost of producing oils including the entry of new technologies to maximize oil recovery at lower costs.

Barriers

Environmental safety. It has been long proven that drilling wells, fluids and the oil itself affects the environment—especially marine water in offshore drills—that is why it is important to ensure that fluids used in drilling should be biodegradable and safe [20].

High cost of nanomaterials. The cost of graphene sheets and the additives to disperse or functionalize them are particularly high. However, in the case of Graphene Nanochem, it has its own proprietary patented manufacturing techniques for graphene as well as its own patented dispersion additives called PlatSperse™.

Enablers

Acceptable results by independent testing regulatory bodies. The PlatDrill™ Series were tested for 9 months by the Council of Scientific and Industrial Research—National Institute of Oceanography, India, and was found out to be 25% more biodegradable than the standard specified by the Organization for Economic Cooperation and Development guidelines and, therefore, was considered environmentally safe and non-toxic in addition to an agreement with a major oil company to test its product.

Joint ventures. Graphene Nanochem has signed a number of joint ventures to enable it to fulfill the commercialization of its products. The company made a joint venture with Scomi Oil tools to enhance its manufacturing and commercialization capabilities of oilfield chemicals and drilling fluids to be able to serve Scomi's wide customer base in 22 countries.

Signing agreements with big oil companies. Graphene Nanochem signed an agreement with Fire Creek Resources Ltd, an oil drilling company for market expansion in North America. It also made agreements with Petroleum Pipe Company for prototyping which makes pipes and casings using high performance materials.

Specialization and experience. Graphene Nanochem has been originally in the field of nanotechnology since 2006 under the name of Biofutures which was changed to Graphene Nanochem in 2012. It has several manufacturing facilities in Malaysia that are specialized in the Bottom-Up synthesis of graphene, dispersion of graphene and the making of finished products to suit the different applications.

Patents. The company has a number of approved patents in graphene synthesis and application.

Targeting several sectors. Graphene Nanochem used its expertise, not only target the oil drilling companies, but also to target water treatment in the oil sector by cooperating with HWV Technologies that specialize in water decontamination and remediation for a variety of applications. Graphene Nanochem has also collaborated with Sync R&D (an engineering company) to develop graphene enhanced lithium batteries to drive electric buses used in Malaysia in a Malaysian Electric Bus Program.

Impact

Effective products. Due to its experience and successfulness, the company was capable of selling several products including PlatDrill™ Series, PlatQ™ Series, PlatSperse™, PlatAmber™ and PlatClear™ not only in the UK but around the globe.

Better well penetration. Such drilling fluids have enabled better penetration of drilling wells while protecting the drilling equipment and allowing better extraction of the oil from such wells due to their controllable electro rheological properties.

Reduce friction and corrosion. They were also capable of reducing friction and heat, impeding corrosion, maintaining a stable rheology, reducing torque and drag, increasing extreme pressure lubricating properties in drilling fluids and high stress environments as well as cleaning, conditioning and strengthening of the well surfaces.

What's Next?

Graphene Nanochem has potential to use its technology and experience in graphene dispersion to develop and make conductive inks and nanocomposites to produce products in areas where it can target other markets as well.

SOLAR BOTANIC LTD.

Electricity generating tree!

Solar Botanic designed an artificial tree with artificial leaves that are made of tiny solar cells and that is capable of generating electricity.

Innovation

- Solar Botanic Ltd has developed a new renewable energy conversion method (an artificial tree) that uses a combination of nanotechnology and biomimetics to achieve its goal.

Nanotechnology

- The company uses Piezoelectric nanoelements converting movement due to wind into electricity. It also uses nanoantennas on the leaves acting like nano solar cells that capture the sun's energy.

Results

- The new technology and design has helped achieve a fifty percent increase in the efficiency of conversion of solar and wind energy to electrical energy. It also reduces the cost per KWh and offers a broad range of applications (grid and off-grid).

Drivers

Increasing global demands. The global demands for energy are rising as the population increases which puts a lot of pressure on generating energy that majorly comes from the non-renewable fossil fuels.

Pollution. The major sources of energy till now is the combustion of fossil fuels which generates a lot of carbon dioxide, carbon monoxide and nitrogen oxides that are toxic and contribute to global warming. In addition, other emissions of soot particles and other pollutants are produced which were found to be major causes of cancer, Alzheimer's and many spreading diseases nowadays.

Higher energy harvesting efficiency. Until now, the energy harvesting efficiency of solar panels and windmills is low and there is increasing research nowadays to increase the efficiency of such clean potential energy sources.

Barriers

Technology. The technology of making piezoelectric ceramic nanoribbons and nanosolar cells imprinted on the rubber leaves is complex and requires great accuracy.

Design. The design of leaves similar to those found in nature; to be able to place the leaves in the right positions for optimum energy capture from wind as well as light are some of the major limitations. Another barrier is creating the right design of leaves and trees to help them get rid of humidity and dirt falling on them as they occur.

Maintenance and durability. The ability of rubber and plastic to withstand harsh environmental conditions without loss of their functional efficiency is a very important aspect that should be considered.

Enablers

Hybrid energy system. It converts the energy from the sun, wind and rain into electrical energy which increase its total output of energy and, hence, its efficiency.

Looks good aesthetically. Unlike the conventional solar panels or the wind turbines that look big and artificial, this new energy conversion system looks just like a tree with leaves, branches and a trunk which make it look beautiful as if it were a real tree. It provides shade and contributes to the visual beauty of the place.

Light in weight. The leaves and branches are made from light rubber and plastic so they become sensitive to motion of wind. They are, however, very robust which decreases the need for maintenance. The light weight makes them easy to install.

Low cost. They are made from recyclable materials which reduces their costs and makes them environmentally friendly.

Impact

Due to its one of a kind technology and biomimicry, Solar Botanic is the leading company in harvesting combined energy from wind and light.

What's Next?

The company is making different designs and sizes of leaves to fit different circumstances as well as optimize the designs to maximize energy capture potential. It is also seeking cheaper manufacturing methods to make such trees even lower in price.

SPIRE SOLAR BIPV

Chicago: *the tale of a green city*.

Spire Solar supplied the city of Chicago with innovative and cost effective solar energy systems to meet Richard Daley's (Mayor of Chicago) ambitions in turning Chicago to the most environmentally friendly city in the US nation.

Innovation

- Spire solar uses nanostructured materials to fabricate more efficient Building Integrated Photovoltaics (BIPV). BIPV can be integrated into façade elements like windows, awnings, or even as curtain wall systems, especially that the fabrication process enables their solar cells to be available in different colors providing architects with different options for better aesthetics [21].

Nanotechnology

- They use nanostructured materials (including quantum dots) that improves the efficiency of BIPV. BIPV are photovoltaic cells that can be integrated into buildings to replace conventional construction materials.

Results

- Chicago implemented the largest expansive solar project in the US where around 28 independent solar power systems were installed in museums, schools, parks, and courthouses generating, collectively, over one million watts of power and displacing around 25 million pounds of carbon emissions. According to LEED criteria [22], Chicago was named the number 1 green city by the business facilities magazine.

Drivers

Nano contributions. Nanotechnology is expected to show great improvements in solar power generation, manifesting in the great enhancements in solar power, efficiency, increased battery storage capacity, lifetime, and reduction of costs.

Energy crisis. The global primary energy demand is increasing: the global increase in carbon emissions are driving the need for efficient sustainable sources of energy.

Visionary mayor. Richard M. Daley, Mayor of Chicago, had a green vision to remediate Chicago's brownfields, and to turn them into solar energy powered plants to generate clean electricity that can help remediate other brownfields into bright ones [23].

BIPVs are the near future. According to the department of energy, over 50 % of the US electricity needs can be supplied by BIPV [24].

Barriers

Conversion efficiency. Nano-based solar panels are still less efficient than silicon-based ones, with a solar conversion efficiency of 6–13 % compared to 20 % in silicon-based ones [25].

Toxicity and scarcity. The nanomaterials used in photovoltaics PV include gallium, indium and other heavy metals which may pose toxicity risks to human health and the environment. In addition to that, their shortages that require strong intervention and efficient recycle processes to recover them from disposed panels make them extremely costly.

Opposing opinions. Some opponents are against governmental support of solar cells based on rare or scarce metals because in the long term they won't be the optimum solution for the energy crisis.

Unprecedented project. A project of such magnitude was the first of its kind. There were many regulatory, financial, technical hurdles to overcome, because there was no clear protocol to manage a project that never existed.

Enablers

Nearby factory. Spire Solar set up a manufacturing facility on Chicago's West Side to support the local participating sites in Chicago with PV panels.

Open innovation. Solar project experts from Spire Solar worked in collaboration with the local government Commonwealth Edison (ComEd: the city's privately held electricity provider) and organized labor. Such big collaborations between different parties in Chicago made the Chicago green project real and successful [23].

Business model. Spire solar Chicago adopted a successful business model to mitigate the risks of being first entrants in the region where it had agreements on guaranteed purchases of its products from the city and ComEd.

Photovoltaic Incentive Program (*PIP*). In an attempt to encourage the wide installation of PV in Chicago, ComEd, in partnership with Spire Solar Chicago, offered local residents a rebate on the installation of PV in residential or commercial buildings, in addition to other governmental incentives on newly installed PV.

Impact

Financial. Spire solar BIPV reduces the cost of integrating conventional PV by offsetting the costs of the replaced construction materials.

Economical. The Chicago solar project didn't only remediate brownfields into Brightfields, it also created new job opportunities due to the new established Spire Solar facility on a Brightfield. In addition, the produced solar systems could generate revenues that can be used to remediate other brownfields.

Social. The project was so enormous and successful that it served to increase the public awareness about solar energy and its benefits, in addition to many cities working on replicating Chicago's solar initiative.

LEED certificate. One of the participating buildings in the Chicago project is the Chicago Center for Green Technology, which was awarded the U.S. green building council LEED platinum rating for its rooftop PV installations.

What's Next?

Chicago's commitment to its green initiative in partnership with Spire Solar will help the company achieve 25 % reduction in carbon emissions by 2020 and 80 % by 2050 [26].

ROYAL DSM KHEPRICOAT

Capture more sun. Break a record.

A new anti-reflective coating that can reduce solar energy losses helped an innovative solar energy module break a record.

Innovation

- A nanoporous coating where the pores are in a closed system protecting the nanomaterials' features against hydrolysis, dust, and even extreme weather conditions.

Nanotechnology

- KhepriCoat's anti-reflective glass coating is a thin nanoporous coating layer 100–150 nm that is applied to the cover glass of solar modules (PV and solar thermal modules) to decrease the reflection of sunlight [27]. The nanoporous coating is made of core shell particles that have a polymer core and silica based shell. The particles are bound together by modified silica binder. When the coating is applied on glass, the binder forms a solid network, and upon thermal treatment, the polymeric core is removed leaving high internal porosity [28].

Results

- The Norwegian companies Renewable Energy Corporation (REC) and Energy Research Centre of the Netherlands (ECN) applied KhepriCoat to develop the world's first multicrystalline solar panel to achieve 17 % conversion efficiency, which was tested under standard European testings done by the European Solar Test Installation [29].

Drivers

Demand pull. The global demand for sustainable energies is increasing as solar systems become the primary focus.

Energy crisis. More efficient solar systems will make critical contributions to solving the energy crisis.

Government push. Governmental and public pressures are forcing companies to work hard on improving the cost-performance ratio of solar modules.

Competitive market. Solar energy companies are continuously competing to break a record and improve the solar conversion efficiencies of their solar energy modules.

Barriers

Overcome cost barriers. Solar panels are still an expensive technology as anti-reflective coatings need to be single layer to remain cost effective. Therefore, premium price has to be justified by optimum efficiency.

Successful integration. KhepriCoat is not one fit for all solar panels; its successful integration with REC and ECN solar panels was a challenge.

Technical conflict. It is particularly challenging to solve the tradeoff between optical and mechanical performance of anti-reflective coatings (AR) because conventional AR coatings are made of silica nanoparticles bound together by a binder with spaces in between acting as nanopores. Too much of the binder reduces the nanoporosity and compromises the optical performance. On the other hand, too little binder compromises the mechanical performance [28].

Enablers

Competence enhancing. The antireflective coating can be applied by a range of conventional deposition techniques, thus, eliminating the need for totally new machinery.

Dedicated partner. Long before the record achievement, ECN had been interested to work with DSM to explore the potential added value of its new anti-reflective coating.

Controllable features. The coating thickness and optical property can be tailored according to the spectral features of the newly developed solar modules.

Impact

Global record. The coatings showed significant contributions in improving the energy output from REC and ECN multicrystalline solar photovoltaic (PV) panels. They were the first panels in the world to achieve 17 % conversion efficiency.

Operational. The light transmission of coated solar modules increased up to 4 %, thus increasing their cost efficiency, and reducing the costs of the energy itself [30].

Certification. Solar modules coated with the innovative nanoporous coating could meet the standard durability criteria and receive Technischer Überwachungsverein (TÜV) certification [28].

What's Next?

Royal DSM is currently involved in developing long-lasting anti-soiling coatings to reduce the maintenance required for optimal module productivity, and for anti-reflective coatings specifically designed for thin film applications. Moreover, Royal DSM opened a manufacturing plant in Chemelot in the Netherlands for its patented Khepricoat to meet the growing needs of the solar modules market.

PHOENIX MOTORS

Plug in: Recharge your car in less than 10 min!

California based Phoenix motors uses Altairnano's nanosafe batteries in its electric vehicles.

Innovation

- Conventional lithium ion batteries (LIB) use graphite which decomposes by time. Altairnano replaced graphite with a patented nanomaterial that resists the decomposition process.

Nanotechnology

- The company used nanostructured lithium titanate: It has a large surface area to facilitate electric-charge transfer.

Results

- Premier fleet companies in California demanded orders for over 500 Nanosafe Phoenix powered electric vehicles on the first year of their introduction, creating a huge potential for the future of these vehicles in the US market.

Drivers

Technology push. Nanotechnology can have a great contribution in improving the performance and durability of conventional LIB and reducing their manufacturing costs.

Regulatory push. The state of California launched the Zero Emission Vehicle program (ZEV) to reduce the increasing air pollution. The program enforces auto manufacturers to produce a certain percentage of zero emission vehicles as electric and hybrid vehicles.

Economic concerns. The US government is encouraging the widespread introduction of high performance electric cars because of the high dependence of the state on foreign oil and the rising oil prices.

Barriers

Limited capacity. The initial numbers of launched Phoenix cars is limited by the limited manufacturing capacity of Altairnano which translates into slow cash flow which is needed to expand its manufacturing capacity.

Public misconception. The public perceives the electric vehicles as low performance ones compared to gasoline vehicles. This slows down the wide market adoption rate of electric vehicles.

Technology expenses. Although LIB are the perfect chemistry batteries for transportation vehicles, they are also the most expensive in terms of their development and improvement.

Enablers

Marketing strategy. The state of California has more than 200,000 fleet vehicles, thus Phoenix attempted to initially target the growing fleet vehicles market in California.

Premium partner. Altairnano is one of the top nanotechnology companies that has spent several years working to improve the performance of LIB and has, successfully, been the first to replace graphite with nanotitanate.

Unmet market needs. Phoenix spotted an excellent opportunity to play a leading role in developing high performance electric cars in a market that is at its infancy and that has a strong potential to grow.

Impact

Green leader. With Nanosafe technology, Phoenix SUV and SUT electric vehicles have been capable of meeting California's Type III ZEV requirements.

High performance. The nanotechnology implemented in Nanosafe batteries produced high performance vehicles that can be recharged in less than 10 min using a 480 V circuit compared to several hours in conventional electric vehicles.

Market expansion. Phoenix's high performance electric vehicles attracted deals from outside California where the state of Hawaii signed deals to introduce the Phoenix electric vehicles into its market in an attempt to secure an increased dependence on renewable energy.

What's Next?

Market acceptance of electric vehicles is increasing in the US especially that nanotechnology advances can potentially enhance the performance cost aspects. Therefore, strategies are investigated to create a network of high power electric outlets to facilitate long journeys by electric vehicles.

References

1. Yang W, Li J, Zhang Y, Huang P-K, Lu T-C, Kuo H-C, Li S, Yang X, Chen H, Liu D, Kang J (2014) High density GaN/AlN quantum dots for deep UV LED with high quantum efficiency and temperature stability. Sci Rep 4:5166
2. Minot ED, Kelkensberg FF, Van Kouwen M, van Dam JA, Kouwenhoven LP, Zwiller V, Borgström MT, Wunnicke O, Verheijen MA, Bakkers EPAM (2007) Single quantum dot nanowire LEDs. Nano Lett 7(2):367–371
3. SoukhtSaraee H, Jafarmadar S, Taghavifar H, Ashrafi SJ (2015) Reduction of emissions and fuel consumption in a compression ignition engine using nanoparticles. Int J Environ Sci Technol 12(7):2245–2252
4. Ponmani S, Nagarajan R, Sangwai J (2013) Applications of nanotechnology for upstream oil and gas industry. J Nano Res 24:7
5. Abdel-Mottaleb MSA, Byrne JA, Chakarov D (2011) Nanotechnology and solar energy. Int J Photoenergy 2011:194146
6. Simon P, Gogotsi Y (2008) Materials for electrochemical capacitors. Nat Mater 7(11):845–854
7. Panero S, Scrosati B, Wachtler M, Croce F (2004) Nanotechnology for the progress of lithium batteries R&D. J Power Sources 129(1):90–95
8. Zhu M, Lu Y, Ouyang L, Wang H (2013) Thermodynamic tuning of Mg-based hydrogen storage alloys: a review. Materials 6:10
9. Schlapbach L, Züttel A (2001) Hydrogen-storage materials for mobile applications. Nature 414:353–358
10. Sandí G (2004) Hydrogen storage and its limitations. Electrochem Soc Interface 13:40–44
11. Bououdina M, Grant D, Walker G (2006) Review on hydrogen absorbing materials—structure, microstructure, and thermodynamic properties. Int J Hydrog Energy 31:177–182
12. Kohno T, Yoshida H, Kanda M (2004) Hydrogen storage properties of La $(Ni0.9M0.1)_3$ alloys. J Alloy Compd 363:254–257
13. Stetson N (2012) Hydrogen storage sub-program opverview. WA: FY 2012 Annual Progress Report, IV-3–IV-10
14. Jeon K-J, Moon HR, Ruminski AM, Jiang B, Kisielowski C, Bardhan R, Urban JJ (2011) Air-stable magnesium nanocomposites provide rapid and high-capacity hydrogen storage without using heavy-metal catalysts. Nat Mater 10:286–290
15. Barkhordarian G, Klassen T, Bormann R (2006) Catalytic mechanism of transition-metal compounds on Mg hydrogen sorption reaction. J Phys Chem B 110:11020–11024
16. Garnett E, Yang P (2010) Light trapping in silicon nanowire solar cells. Nano Lett 10(3):1082
17. (n.d.) http://www.iea.org/publications/freepublications/. Accessed 5 May 2015
18. Kallman M (2008) Air pollution's causes, consequences and solutions. World Resources Institute

19. Kasiralvalad E (2014) The great potential of nanomaterials in drilling and drilling fluid applications. Int J Nano Dimens 5(5):463–471
20. Olsgard F, Gray JS (1995) A comprehensive analysis of the effects of offshore oil and gas exploration and production on the benthic communities of the Norwegian continental shelf. Mar Ecol Prog Ser 122:277–306
21. Elvin G (2007) Nanotechnology for green building. Green Technology Forum
22. Chicago bright fields: a success story. (n.d.) http://infohouse.p2ric.org/ref/17/16869.pdf. Accessed 1 Aug 2014
23. Martin G (2003) Renewable energy gets the "Green" light in Chicago. IEEE Power and Energy Magazine, p 34–39
24. (n.d.) http://nanobite.blogspot.com.eg/. Accessed 1 Aug 2014
25. Illuminato I, Miller G (2010) Nanotechnology, climate and energy: over-heated promises and hot air? Friends of the Earth. http://www.foe.co.uk/sites/default/files/downloads/nanotechnology_climate.pdf. Accessed 1 Aug 2014
26. (2010) Chicago climate action plan. http://elpc.org/wp-content/uploads/2010/10/City-REWG.FINAL-REPORT.September-2010.pdf. Accessed 1 Aug 2014
27. (n.d.) http://www.renewableenergyfocus.com/view/32127/royal-dsm-opens-plant-to-make-anti-reflective-coating/. Accessed 3 Aug 2014
28. (n.d.) http://www.slideshare.net/Khepricoat/khepri-coat-brochure-v1-2011. Accessed 3 Aug 2014
29. Tjengdrawira C, Lamers MWPE, Bennett IJ, de Jong PC (2010) World first 17% efficient multi-crystalline silicon module. Paper presented at the 35th IEEE Photovoltaic Specialist Conference, Honolulu, HI, 20–25 June 2010. Energy research Centre of the Netherlands, Netherlands
30. DSM opens new manufacturing plant for innovative anti-reflective coating (2013) http://www.dsm.com/corporate/media/informationcenter-news/2013/04/2013-04-22-dsm-opens-new-manufacturing-plant-for-innovative-anti-reflective-coating.html. Accessed 3 Aug 2014

Chapter 5
Environment and Remediation Applications

Environmental pollution is an increasingly alarming issue that has recently drawn a lot of attention. Pollution has resulted in the outbreak of a lot of diseases including cancers, Alzheimer's, chronic lung allergies and numerous infectious diseases worldwide. It has resulted in the loss of a lot of ecosystems and in the depletion of biodiversity with the extinction of a great number of species of plants, insects and animals. Pollution has also resulted in the global warming effect with significant effects on the weather and worldwide climate. This has lead scientists to develop recycling, remediation and environment cleaning methods to help reduce such disastrous environmental effects. Since nanotechnology is a new and fascinating technology that offers small size, low cost and high efficiency, it has drawn attention to its potential usefulness in the fields of pollutant sensing, remediation, contaminant removals and recycling. Many companies have designed nanomaterials or nanotechnological processes that offer more efficient environmental solutions than conventional processes and have succeeded in reaching worldwide markets and undertaking global remediation projects.

Introduction

An increasing interest and a large amount of research in remediation and environmental cleanup has been emerging around the world in the past few years. In 2009, the American Recovery and Reinvestment Act allocated approximately 1 billion dollars to the US Environmental Protection Agency to be spent on remediation projects [1]. The use of nanotechnology and nanomaterials in remediation has drawn a lot of attention in the past 5 years due to their small size, large surface area, high selectivity and reduced cost. According to a report by the European NanoRem project, there are about 70 nanoremediation

M. Anis et al., *Nanovate*, DOI 10.1007/978-3-319-44863-3_5

projects worldwide and the number is increasing [2]. Nanotechnology is of use in environmental cleanup in three key areas including: environmental sensing and monitoring (including air, water and soil), nanoremediation and contaminant removal as well as pollution prevention (by helping to make sustainable or environmentally green products) [3, 4]. Sensoring and monitoring include solid state gas sensoring for quick real time detection of contaminants in air. These are usually made of thin films of metal oxide nanocrystals that are much cheaper than ordinary detection methods [5]. Nanosensors for water monitoring have also been designed that are capable of detection limits of less than a μg/L capable of detecting a variety of compounds including antibiotics, hormones, pesticides, metals and many other contaminants [6]. In remediation, the use of very small nanosized materials that can adsorb or degrade contaminants are injected in the contamination site where it performs its decontamination function in situ giving it an advantage over the conventional remediation methods.

Cases

NANOSTRUCK, INC.

Gold from wastes!

Nanostruck was able to device a new method of recovering metals from wastes.

Innovatxion

- New recovery methods for platinum, gold, silver, uranium and other precious metals from mining wastewater to prevent it from seeping to drinking water.

Nanotechnology

- Nanostructured copolymer molecular sponges made from naturally occurring polymers like chitosan are programmed to filter and extract such metals as gold, platinum and many other metals using physio-mechanical capture principles.

Results

- This technology has enabled the efficient and economic removal of toxic pollutants in the wastewaters of many mining processes enabling the reuse of water in agriculture and irrigation as well as household and drinking water. In addition, the Nanomet® technology has allowed the efficient collection and isolation of precious metals that can be recycled for reuse in the various industrial purposes. This has made NanoStruck Inc. an important cooperator of some of the largest sanitation projects around the globe as well as aiding it in coordinating with several industries to help them recycle and reuse their byproducts.

Drivers

Global water shortage problems. The world's population is in constant growth predicted to reach 6.3 billion people in 2050 [7]. This signifies an alarming increase in demand for water for agriculture, domestic and, more importantly, drinking water. Presently, 64 billion cubic meters of fresh water are being consumed yearly according to the World Water Development Report in 2014 [8] and in many areas, there are water shortages due to polluted and/or dried up rivers resulting from climate changes.

The worldwide contamination of waters especially in urbanized areas. Urbanization and industrialization has led to increasing pollution levels in air and water with heavy metals. The increased pollution levels in water due to wastewater disposal in rivers or underground without treatment has caused the spreading of diseases and toxicity incidents among the people which has driven a lot of countries to set regulations as to the treatment of waste waters from factories before disposal [9].

The high cost of other conventional remediation methods. Nanoremediation offers cost solutions lower than the conventional remediation methods for three main reasons. One is that they allow in situ treatment without the need to pump the water from its site to be treated elsewhere in a plant. The second is that the small size enables them to be added to the wastewater site. Thirdly their large reacting surface areas enable the accomplishment of the decontamination job for much less material than that used in conventional methods [10].

The need to secure long-term precious metal resource availability. Precious metals are rare and have high value stored in them. Gold for example is used to balance the economies of many countries. The saving of leached precious metals and their reuse can help save a lot of costs lost in several manufacturing processes.

Barriers

Complexity of the technology. The designing and making of nanomaterials is a complex process that requires a lot of expertise which is difficult to find as this is still a new and young science.

Safety of the nanomaterials used. The exact toxicity profiles of nanomaterials have not been fully developed yet; how they interact with human cells, whether they are carcinogenic and their interaction with other soil and environment components to form toxic products [11].

Clumping of nanomaterials. A disadvantage of such small sized materials is that they have high surface energy due to their size and large surface area which makes them more susceptible to clumping and reducing the efficiency of distributing them in treatment sites [12].

Lack of supporting and regulatory laws. Due to the still young science of nanotechnology as well as the lack of sufficient testing regarding the safety of the use of nanomaterials and the presence of many contradictory reports, the development of government and other regulatory policies has not been fully established yet in most countries [13].

Enablers

R&D. Nanostruck, Inc. has carried research in the field of nanoremediation for several years and has acquired several patented technologies and products including sand filter, centrifugal separator, electronic precipitator, cleaning universal strainer, chitosan nano copolymers, mine dam water treatment, nanosan toilets, landfill leachate process and extraction of gold and other metals. It has collaborated with high caliber scientists in the University of Saskatchewan in Canada to be able to produce successful nanoproducts.

A successful cost efficient product. The Company uses the Environmental Protection Agency (EPA) and World Health Organization (WHO) guidelines as standards for water quality and safety when designing its nanoremediation products.

Impact

Used by many water and sewage treatment plants. Due to its cost effective product, Nanostruck technologies has been incorporated in several plants in Canada and Mexico.

Being involved in the precious metal recovery project in South Africa. The company built a plant in South Africa to recycle precious metals.

Using the technology to create household sanitation platforms that can be used in rural areas. Using its technology, Nanostruck was capable of making a sanitation system for use in homes in rural areas at low cost for the ordinary end user.

What's Next?

Using its new and successful technology, the company is negotiating over large sanitation projects in South Africa as well as joining other industries, like jewelry manufacturers, to help them recycle and reuse any leached gold or platinum in their manufacturing processes.

HYPERCAT ACP™

Emission reduction.

Hypercat devised a new sponge that degrades emissions.

Innovation

- New cheaper and more efficient emission control catalysts for engines and generators with less use of Platinum Group Metals (PGM) such as ruthenium, rhodium, palladium, osmium, iridium, and platinum.

Nanotechnology

- Nanoceramic composites and nanoalloys in the forms of filters, foams, sponges or just substrates in which nanoparticles of noble metals like platinum or palladium are immobilized in a similar manner as shown in Fig. 5.1 are designed with specialized properties to act as catalysts for degrading compounds in emissions of engines and generators.

Results

- It was capable of fabricating efficient catalysts that are much cheaper and more efficient than conventional ones and that can be designed to fit a variety of shapes and sizes. Such catalysts are capable of removing three times more harmful nitric oxide gas than the conventional catalysts.

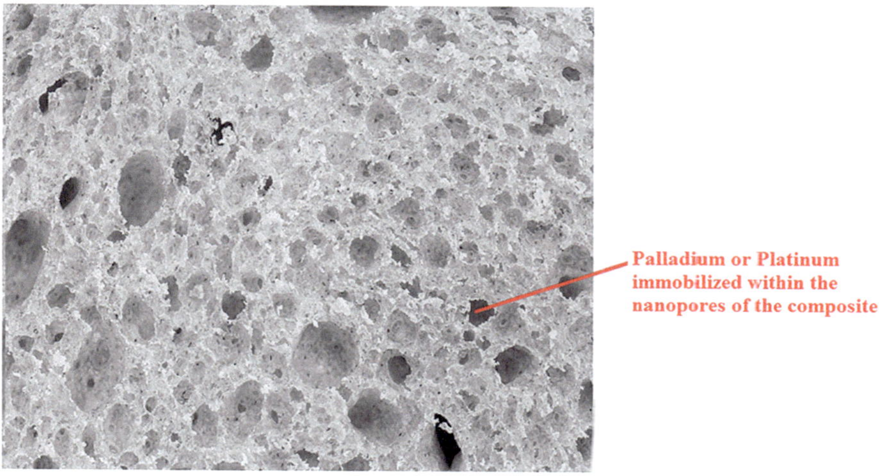

Fig. 5.1 Pd immobilized in Cordierite (Magnesium Iron Aluminium Cyclosilicate) ceramic Sponge

Drivers

Air pollution-major environmental problem. According to the Blacksmith Institute report [14]: World's Worst Polluted Places, air pollution is listed as the most toxic and worst pollution problem. Additionally, the world health organization [15] reported that air pollution has caused the death of 7 million people around the globe in 2012.

The low efficiency and high cost of conventional catalytic converters to remove nitric oxide. In the nanosize, range materials have a much larger reactive surface area than those not in the nano scale making them more efficient in catalysis while, at the same time, using less material than if they were in their big size [16].

Spreading of lung diseases and other pollution related diseases. Recently, a lot of research has been carried out on the effects of air pollution on populations and their health.

It was found that air pollution was associated with lung, cardiovascular, neural diseases and that it was associated with mortality, hospital admissions and malformation of children at birth [17].

The new direction of the auto industry. The automotive industry has changed its strategic direction along the past few years switching to more sustainable and greener fuel options as seen by the immense research put by these companies to make the hybrid fuel cells, hydrogen cells and the electric car.

Reduction of emission suffocating odors. Not only are the fumes produced by combustion harmful and contribute to pollution, but also they reek.

Barriers

Research and development. The use of nanotechnology and the accurate custom design of the catalytic converter in the right way to fit its purpose are challenging barriers. The company, however, has built experience since 2005 in the art of making nanocatalysts and designing the catalytic converters. It utilizes the help of chemists as well as mechanical engineers to get its job done as well as allocating a 9000 square foot facility just for research.

The probability of catalysts to become deactivated at high temperature and lose their functions. Complexing and immobilization of the nanoparticle catalysts in a composite helps to protect them against the actions of high temperature.

Enablers

Global and governmental restrictions on emissions. Strict regulations concerning emissions from factories, power plants and automotives have been set by the Environmental

Protection Agency (EPA) through the Clean Air Act regulating the emissions of oxides-of-nitrogen (NOx), oxides-of-sulfur (SOx) emissions, heavy metals (Hg, etc.) and carbon dioxide [18].

The advancement of technology in catalysis especially nanocatalysis. The advancing technologies in surface chemistry and nanocatalysis has enabled the formation of emission control catalysts with greater efficiency and lower cost [19].

Impact

A cost efficient product that helped the company reach a variety of sectors including the air and climate, energy, food and beverage, manufacturing, oil, gas and refineries and printing industries. In coffee manufacturing, overwhelming odors are released which has driven several companies to use the Hypercat technology as it is capable of removing volatile organic compounds. The production of suffocating odors has brought a lot of trouble to manufacturers, so much so that they were threatened to be shut down by governments; Sriracha Plant in Southern California is a relevant example [19].

Making joint venture with big companies like Nanostellar, Inc. Nanostellar, Inc.: a large manufacturing company making materials for NASA was lured by Hypercat expertise in catalytic emissions control and its products to join venture with it.

What's Next?

Hypercat ACP has a future potential through reaching out to various large global companies like car companies and those who are interested in sustainable and environmentally friendly emissions to sell them its technology.

NANO ENGINEERED APPLICATIONS, INC.

A smelling nose chip!

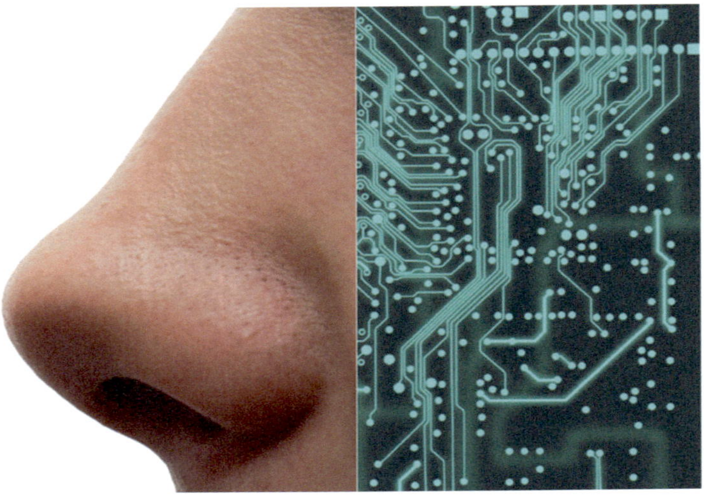

Nano Engineered Applications devised a novel gas sensor.

Innovation

- A tiny gas sensor in the form of a chip that allows real-time detection of several chemicals in the air at very low concentrations (at the parts-per-billion (ppb) level) without needing too much battery power.

Nanotechnology

- Carbon nanotubes are used on the chips that are functionalized with different sensing molecules, which is specific to different chemicals, allowing the simultaneous detection of these chemicals.

Results

- Nano Engineered Applications, Inc. is continuously developing a device of important applications in the agriculture (to detect pesticides), industry (gas leaks, combustion emissions), homeland security (bioterrorism) and the military fields (to detect chemical warfare agents). It can also be used for Air Quality Monitoring, Indoor and Outdoor Public Health Monitoring, Food Safety and Food Supply Chain Monitoring and even in early disease detection.

Drivers

The need to monitor potentially hazardous gases. Potentially hazardous gases are continuously produced across a wide range of fields and industries including factories, combustion processes, household processes producing gases such as methane, carbon monoxide and carbon dioxide. The harmful effect of these gases on the human health have become widely known and discovered.

The need for lightweight sensors that can be incorporated in an application without too many overheads. The availability of small sensors that can be incorporated into processes without changing their whole architecture is extremely necessary.

Rapid and real-time sensing to enable monitoring of processes. Old conventional sensors did not have real time sensing as with the new ones that can sense several molecules simultaneously and send signals as they sense their presence or absence.

The need for low power detection systems. The development of detection systems that require low battery power has enabled such detectors to be used remotely for a long time without needing replacement.

Barriers

Research and development. The research and experience required to design such small sensors.

The cost of fabricating such sensors with high technology. Technologies such as aligning carbon nanotubes or the designing of solid state electric sensing using nanomaterials are expensive processes and would require high initial costs.

Enablers

Partnerships and collaborations. Nano Engineered Applications, Inc. is a part of ieCrowd™ Company, an innovation company built to foster new technologies to boost economy and used to raise funds. It also has the support of the University of California, Riverside and National Institutes of Health.

Research experience. The company is formed of a group of scientists who are experienced in solid state electrical sensing at the University of California.

Impact

Competitive products in a new emerging market. Due to the company's state of the art patent technology, it is capable of entering a still young market with a competitive product making it a first mover and enabling it to monopolize the market later on.

What's Next?

The next step for the company with its state of the art products would be to commercialize the products globally and use such sensors to be incorporated in wearable or handheld mobile devices such as wireless health monitors and smart watches and to target an even wider consumer base.

DAIS ANALYTIC CORPORATION: CONSERV™

An Energy Recovery Ventilation (ERV).

Dais Analytic designed a new ventilation nanocomposite system that allows better recovery of energy.

Innovation

- Designing a new energy recovery ventilation system that reduces CO_2 emissions in the outside environment and is more efficient.

Nanotechnology

- The nanotechnology used is a highly hydrophilic nanocomposite polymer membrane (derived from sulfonation of styrene-containing block copolymers) that promotes ionic conductivity, transport of water (in the form of vapor) and exchange of energy along with the moisture in the form of latent heat.

Results

- Through using nanotechnology, Dais Analytic Corporation was capable of making a more efficient heat exchange and energy recovery fixed plate ventilator. The ConsERV™ ventilator allowed the exchange of 71% sensible heat (like other fixed plate exchangers) and 64% latent heat through moisture exchange through the hydrophilic nano-designed membrane (unlike other fixed plate exchangers).

Drivers

An increasing demand on energy due to the increasing population. According to the Energy Information Administration [21], the world energy consumption will increase by 56% between 2010 and 2040. This has lead countries around the globe to research ways to conserve energy and minimize its consumption. The energy recovery ventilator by Dais allows the treatment of the incoming air with the exhaust room air to maintain the heat inside and thus minimize energy usage.

Air pollution and global warming. It is well known that the release of CO_2, CO and NO_x from consumption of fossil fuels for heating pollute the air and are deleterious to the human health. Additionally, these gases, especially CO_2, contribute to the greenhouse effect and global warming phenomenon. According to the Intergovernmental Panel on Climate Change (2014), it was reported that during the twenty-first century the temperature would rise a further 0.3–1.7 °C (0.5–3.1 °F) for their lowest emissions scenario using stringent mitigation and 2.6–4.8 °C (4.7–8.6 °F) for their highest.

The need for a robust, low maintenance air ventilation system. The ConsERV™ ventilator is a fixed plate ventilator with no moving parts and less maintenance. It offers exchange of moisture in the form of vapor rather than water which prevents condensation, dripping and the growth of molds.

Barriers

Complexity of the technology. Fabricating an efficient energy recovery ventilator with nanotechnology is not an easy task. However, Dais' extensive research in nanotechnological solutions for a variety of applications has helped it gain expertise and a reliable background.

Competition from other energy recovery ventilators. Most ventilators depend on the transfer of sensible heat only, whereas the ConsERV ventilator introduces the transfer of latent heat with increased efficiency.

Enablers

Research experience. Dais Analytic has been in the field of nanotechnology research since 1999 and was listed in Florida as a nanotechnology polymer materials and processes company.

Collaborations. Due to its outstanding certified products and patented technologies, Dais was able to collaborate with the US Commerce in China for Clean Energy Solutions as well as the US Department of the Navy and the US Department of Energy.

Government grants and awards. Dais was granted an economic incentive award by Pasco County to help it speed its progress in developing its nanotechnological solutions [22].

Impact

Dais Analytic has successfully managed to hold a strong product profile in the market featuring four main products: ConsERV™, NanoAir™, NanoClear™ and NanoCap™. Due to the company's state of art technology, it managed to reduce the costs for ventilation air by as much as 30% and decrease the size of the equipment needed by more than 50%.

What's Next?

Dais can make additional energy conserving products using nanotechnology and incorporate other nanotechnologically enhanced features like filtration systems for different bacteria and pathogens to target different sectors of the market.

GENANO®

Pure nano free air.

Genano has devised a new air purification system that uses ions to kill any microbes in the air.

Innovation

- New high efficiency air purification system that is capable of purifying air from nanosized microbes, viruses and other nanoparticles.

Nanotechnology

- Genano® uses a jet of ions to cause any nanoparticles in the air like bacteria or viruses to stick to adsorbing plates. It also uses carbon nanofilters for removing volatile organic compounds and other odors.

Results

- The technology is able to remove microbes, gases, odors, DNA and nanoparticles up to 3 nm in size.

Drivers

Increasing epidemic diseases like MERS and Ebola viruses. Epidemic disease outbreaks occur every now and then in different areas around the globe e.g. The Middle East Respiratory Disease (MERS) in Saudi Arabia and Ebola virus in West Africa in 2014.

Critical hospital areas like the operation room and the ICU that require purified air. The air in operation and ICU rooms should be exceptionally clean to avoid contaminations during surgery and spread of nosocomial infections that can be hazardous to immunologically compromised patients.

HEPA and ULPA filters do not destroy microbes. These filters are not capable of killing the pathogens they trap. Changing these filters is exceedingly hazardous that it requires the evacuation of the rooms for maintenance to take place. Such filters can remove particles as small as 300 nm but not anything below.

Need for pure air in laboratories to avoid contaminations of tests. In laboratories, especially microbiological and DNA labs, any contamination would entirely alter the results of researches. This necessitates a certain clean air grade.

Air pollution. With the increasing number of populations, number of vehicles and factories, the level of air pollution has been increasing greatly especially in urbanized areas. Such air pollution has drastic effects on the human health including respiratory diseases and cancer.

Increasing sick building syndrome. Nowadays, we spend a lot of time indoors on computers or in front of the television. This causes us to inhale the same unventilated air in the room which usually contains contaminants from furniture, rugs, microbes and moisture-causing products. This induces the development of illnesses such as fatigue and allergies.

Removal of welding fumes and metal processing emissions in industries. In factories, welding and other processes release tiny nanoparticles in the air that, if inhaled by workers, can trigger cancer. This entails an efficient removal system.

Barriers

Maintenance. The maintenance of such a high technology air purifier would present a major barrier to the customer buying the purifier. However, the Genano® purifier offers a no-filter change system that cleans itself through a special washing fluid that is filled in the purifier and the discharged fluids thrown away. It also provides an extensive support and maintenance network all around the globe.

Safety. There have been published reports that any air ionizer would create ozone which is harmful to human health especially in non-ventilated areas [23].

Enablers

Long time experience. Genano has been working in the air purification field since 1998 and has been developing and researching new decontamination methods. This allows it to build a strong base of knowledge and experience, patents, resources, technologies and well developed infrastructure.

Published research results by neutral testing facilities and certifications. The air purification system was tested by a number of independent bodies to prove its efficacy including Metropolilab, VTT Technical Research Centre, Institute of Occupational Health and Epitek Oy for Heating and Air-conditioning in Finland, Ministry of Health in Saudi Arabia and Istanbul University in Turkey.

Major projects. Genano® was involved in combating the MERS that spread in Saudi Arabia in 2012. Its product was installed in a number of hospitals and helped reduce the spread of the illness in the hospitals. The Saudi Arabian Ministry of Health recommended it for installation in the country's hospitals as well. Genano was also involved with Finnzymes Laboratories, part of Thermo Fisher Scientific, to build ISO class 6–7 cleanrooms efficiently, rapidly and at a low cost.

Collaborations. Genano has collaborations with universities and research institutes including: the Finnish Institute of Occupational Health, Ositum, VTT Finland, the KM University of Poland, SGS Germany, Greifswald Institute Germany, LNE France, VITO Belgium, and KF in Saudi Arabia.

Impact

Due to the high quality of its filters, it is recommended by the Ministry of Health for hospitals and laboratories.

What's Next?

Genano can make small sized models that can be used inside transportations like trains, cars and subways.

SELDON TECHNOLOGIES, INC.

Nanomesh filter.

Seldon made a nanopore filter for removing viruses and other nanoparticles from drinking water.

Innovation

- An enhanced water filtration system that can convert any sort of water to clean soft drinking water at an affordable price.

Nanotechnology

- Nanomesh is made up of fused, nano-structured, nonwoven material that combines functionalized carbon nanotubes, activated carbon, and fibers within a supporting substrate sheet material. Using the enormous surface area created by the carbon nanotube-based Nanomesh, it combines size-exclusion filtration and advanced adsorption capabilities resulting in a simple, flow-through filter that maximizes purification at low-pressure drops.

Results

- The technology has enabled Seldon's filtration systems to remove bacteria, viruses, and cysts, like Cryptosporidium and Giardia, and comply to the US EPA Drinking Water Standards (99.9999 percent of bacteria, 99.99 percent of viruses and 99.9 percent of cysts). Furthermore, it aided in reducing sediment, total organic carbon, chlorine, color, bad taste, and odors as well as removing significant levels of harmful chemical contaminants, including lead, cadmium, organophosphates (pesticides and herbicides), disinfection byproducts, and radioactive contaminants such as cesium-137.

Drivers

Polluted drinking water. Water pollution and contaminated drinking waters is an increasing global problem. Water pollution is one of the leading causes of death in the world killing more than 14,000 people daily [24]. Though the results of polluted waters are more evident in developing countries (water borne diseases that cause the death of children), developed countries also suffer. In the United States, 47 % of lakes and 45 % percent of streams are polluted.

Regulations concerning water disposal by factories and industries. The stringent regulations and fines that are imposed on the factories according to the Environment Protection Act forces them to install water purification systems prior to disposal.

Barriers

The presence of other filtration systems. Seldon filters uses nanotechnology that enables efficient decontamination when compared to the conventional filtration systems. Its filters are capable of removing viruses as well as heavy metals, chlorine and odors. An added benefit to Seldon filters is that it makes the filtration systems customized and specialized to the conditions they would be used in, i.e. there are filters for drastic conditions with low water pressure like in remote areas and there are filters for ordinary high pressure tap water.

Safety of using carbon nanotubes. There is still debate as to whether the use of nanomaterials in drinking water or food is safe.

Enablers

Extensive research. Seldon Technologies has been conducting research on the use of carbon nanotubes in filtration for 10 years during which it gained experience and patents. It has a 2500 square feet facility for research work.

Extensive independent testing. The independent testing by parties other than the company has helped validate its filters. Being tested at independent facilities like the National Sanitation Foundation, University of New Hampshire, the international government and military authorized institutes has enabled its products to be recognized.

Targeting several customer sectors. Seldon Technologies has customized its filters to be used by the military as a source of mobile water when they are in remote areas. It also designed filtration systems for humanitarian organizations by providing them with filtration systems that can withstand immensely challenging environments as well as being affordable. The company has also developed the WaterBox™ which is portable and can be used in urgent disastrous situations. It also designed the WaterTap™ for household use. Moreover, Seldon developed the Waterstick™ Ultra: a small water filter with a hand pump used to deliver clean water for people who are biking or in remote areas.

Impact

Due to the cost effective and powerful purification system that it developed, Seldon Technologies was able to enter the American, African, European, Canadian, Australian and Chinese markets. It also received a certificate of compliance granted by the US Army Public Health Center and the National Sanitation Foundation.

What's Next?

Using its patented technology and research facilities, Seldon Technologies can design filtration systems for other non-water applications like oil refinement.

ALUMINASTIC CORPORATION

Cheap functional nanocomposites from garbage.

New high-strength cheap nanocomposites of aluminum from recycled garbage were successfully manufactured by Aluminastic Corporation.

Innovation	•Making new, more ductile and stronger aluminum composites from recycled materials to be used in many fields such as manufacturing, medical and aerospace applications.
Nanotechnology	•Aluminastic was capable of making aluminum nanocomposites with silicon and other materials as well as designing special extrusion equipment for shaping its composites.
Results	•Aluminastic, with its research capabilities and experience, was capable of making cheap recycled nanostructured composites that are more ductile, stronger, more durable and resistive to corrosion, more formable and show excellent finishes. The new composites enable shaping with much lower extrusion force, at lower temperature and at a much faster rate.

Drivers

Increasing garbage. As the population around the globe is increasing, more garbage and wastes are being produced. This waste takes time to disintegrate, thus requiring ways of recycling it rather than disposing of it.

Demand for recycling. The need for recycling has increased over the past years due to the increasing amount of waste (due to increasing population). Increased pollution awareness and increased rules and legislations are also needed. Governments have recently modified their own procurement policies to increase their purchasing of recycled products over non-recycled ones. The Environment Protection Agency in the US has put rules to force the purchase of recycled paper, oil, building insulation and tires.

Need for aluminum composites with better properties. Usually recycled aluminum is less ductile and of lower quality than the new one and there is always a need to make composites and alloys that have higher and much better mechanical properties.

Barriers

Technology. The art of recycling old aluminum products and strengthening them on the nanoscale requires very high technological processes and experience to avoid the segregation of the reinforcing material leading to loss of function and failure.

Standardization. A lot of composites, whether nanostructured or micro structured, are being developed and fabricated each day. However, there are no fixed standards or protocols for the evaluation of their performance. Despite that, Aluminastic are looking for ways to standardize and certify their processes by applying to the National Institute of Standards and Technology, the American Society for Testing and Materials and the American Lightweight Materials Manufacturing Innovation Institute.

Technical difficulty. Although it is known that when silicon is added to aluminum it strengthens it, the process of dispersing the silicon in aluminum is a difficult one and is always subject to segregation.

Enablers

Extensive experimentation. Aluminastic has been experimenting for 12 years to reach the best desired results. It developed its own equipment, materials, procedures and software in the process. It has also acquired many patents in the process.

Testing by independent facilities. Through collaborations and agreements, Aluminastic was capable of testing its products at external entities. The products were tested at the Air Force Research Laboratory at Wright-Patterson AFB in Dayton, Ohio and were validated by Iowa Aluminum, Inc.

High performance product. Aluminastic was capable of making high strength aluminum nanocomposites and of tailoring the properties of the final products to meet the applications they are destined for. The company also made more malleable, more ductile, harder, more wear resistant and lighter aluminum, all dependent on its customers' needs.

Awards received and national recognition. It has received several awards for its successful products and was nationally recognized by CNN for its environmentally green efforts.

Customization. Aluminastic's processing and manufacturing methods allow it to make customized products to suit different applications including aerospace, medical, construction and industrial applications.

Impact

Due to the company's outstanding performance, it was nationally recognized and granted awards. Aluminastic was able to develop and customize a variety of products that can be employed in a number of industries.

What's Next?

The company will be seeking standardization of its products and certifications proving its efficiency in the market to increase consumer acceptance and trust in its technologies and products.

References

1. Karn B, Kuiken T, Otto M (2009) Nanotechnology and in situ remediation: a review of the benefits and potential risks. Environ Health Perspect 117:12
2. Bardos P, Bone B, Daly P, Elliott DW, Jones SE, Lowry GV (2014) A risk/benefit appraisal for the application of nano-scale zero valent iron (nZVI) for the remediation of contaminated sites. *Nanorem*. http://www.nanorem.eu. Accessed 3 Dec 2014
3. Tratnyek PG, Johnson RL (2006) Nanotechnologies for environmental cleanup. Nano Today 1(2):44–48
4. Masciangioli T, Zhang WX (2003) Environmental technologies at the nanoscale. Environ Sci Technol 37:102A–108A
5. Rickerby DG, Morrison M (2007) Nanotechnology and the environment: a European perspective. Sci Technol Adv Mater 8:19–24
6. Zhuiykov S (2012) Solid-state sensors monitoring parameters of water quality for the next generation of wireless sensor networks. Sens Actuators B 161:1–20
7. Bruinsma J (2009) By how much do land, water and crop yields need to increase by 2050? Presented at the food and agriculture organization of the united nations expert meeting: how to feed the world in 2050, Rome, Italy. The resource outlook to 2050
8. World water development report 4 managing water under uncertainty and risks (2014) http://www.unesco.org/new/en/natural-sciences/environment/water/wwap/wwdr/wwdr4-2012. Accessed 5 June 2015
9. Novotny V, Clark D, Griffin RJ, Booth D (2001) Risk based urban watershed management-Integration of water quality and flood control objectives. USEPA/NSF/USDA STAR Watershed Program final report, Institute for Urban Environmental Risk Management, Marquette University, Milwaukee, WI
10. Lowry GV (2007) Nanomaterials for groundwater remediation. In: Wiesner MR, Bottero J (eds) Environmental nanotechnology. McGraw-Hill Companies, New York, pp 297–336
11. Lubick N (2008) Risks of nanotechnology remain uncertain. Environ Sci Technol 42(6):1821–1824
12. Zhang W (2003) Nanoscale iron particles for environmental remediation: an overview. J Nanopart Res 5(3–4):323–332
13. Bardos P, Bone B, Černík M, Elliott DW, Jones S, Merly C (2015) Nanoremediation and international environmental restoration markets. Remed J 25(2):83–94

14. Blacksmith "Reports" (2010) http://www.worstpolluted.org/. Accessed 20 June 2015
15. (n.d.) http://www.who.int/mediacentre/news/releases/2014/air-pollution/en/. Accessed 25 June 2015
16. Chaturvedia S, Davea PN, Shahb NK (2012) Applications of nano-catalyst in new era. J Saudi Chem Soc 16:307–325
17. Brunekreef B, Holgate ST (2002) Air pollution and health. Lancet 360(9341):1233–1242
18. Litz FT, Bianco NM (2010) What to expect from EPA: regulation of greenhouse gas emissions under the clean air act. Envir L Rep 40(5):10480
19. Park JY, Somorjai GAF (2009) Advancing the frontiers in nanocatalysis, biointerfaces, and renewable energy conversion by innovations of surface techniques. J Am Chem Soc 131(46):16589–16605
20. Nash J (2013) In California, a hot topic: sriracha red chili sauce. Pittsburgh Post-Gazette
21. Stocker TF, Qin D, Plattner GK, Tignor MMB, Allen SK, Boschung J et al (2013) Climate change 2013: the physical science basis. Contribution of working group I to the fifth assessment report of the intergovernmental panel on climate change. Cambridge University Press, Cambridge, pp 33–115
22. (n.d.) http://daisanalytic.com/media_news/index.html. Accessed 3 July 2015
23. Britigan N, Alshawa A, Nizkorodov SA (2006) Quantification of ozone levels in indoor environments generated by ionization and ozonolysis air purifiers. Air Waste Manage Assoc 56:601–610
24. West L (2006) World water day: a billion people worldwide lack safe drinking water. *About News*. Accessed 18 July 2015

Chapter 6
Automotive Applications

The market for cars is rapidly growing: in 2008, it was forecasted to double by 2030. With the growth of the car industry, there are also growing negative impacts on the environment as cars are one of the main contributors to the globally rising carbon emissions. This chapter discusses how simple nanomaterials could improve the performances of car engines and at the same time conserve energy. Nanomaterials can be integrated into cars' bodies, tires and engines or used in self-cleaning/self-healing coatings in car parts. The chapter further illustrates how several companies are investing hugely in research and development to add premium features to automobiles and at the same time gain bigger market shares in a market that is known to be highly competitive. The main challenges faced by car manufacturers from a different perspective and how nanotechnology helped them solve some of these challenges will also be tackled.

Introduction

With the expanding global population, the car industry is also experiencing a rapid increase in size and additional economic pressures. According to the United Nations, the market for cars could reach 1.5 billion by 2030, which is approximately double the numbers in 2008 [1]. This rising industry is the second main contributor to the rising carbon emissions worldwide, which is associated with a growing demand for safety and sustainable development. In addition, automotive companies are facing fierce competitions. All of these are driving factors for the companies to invest hugely in R&D with nanotechnology as a premium solution for improving the materials and the performances in the automotive industry. Moreover, automobile companies are currently adding nanofeatures to their cars to differentiate themselves in this highly competitive market.

M. Anis et al., *Nanovate*, DOI 10.1007/978-3-319-44863-3_6

The energy crisis is a global challenge facing the whole world. With emerging trends towards providing sustainable sources of energy, many governments are pressuring car manufacturers to continuously develop and abide by strict regulations that attempt to conserve energy and minimize the harmful environmental impacts associated with car manufacture and use. Nanotechnology in the automotive industry proved to be a cost effective solution. Nanomaterials in car bodies rendered them lighter in weight, more durable, and cheaper. They are expected someday to even render cars self-healing! Moreover, the incorporation of nanoparticles in tires increased their lifetime and enhanced their wear resistance, which increased the safety of passengers. All of these would not only increase the efficiency of engines and reduce fuel consumption, but also would improve the safety of passengers.

Notably, nanotechnology plays an important role in energy conservation by providing more sustainable alternatives. Nanomaterials are incorporated in batteries to improve their performances and decrease fuel consumption and are extensively applied in the form of nanocatalysts to increase the efficiency of engines. This renders the current car market more efficient as well as more environmentally friendly. Extensive research is being undergone to replace conventional engines with fuel cells. These could come in the forms of polymer electrolyte membrane fuel cells or hydrogen fuel cells. Despite this, there are many complexities in their processing and manipulation that delays their wide application in automobiles. Likewise, solar cells are being researched to be integrated into car roofs to act as a clean efficient source of energy. Nanotechnology plays a key role in solving the complexities associated with the development of fuel and solar cells, which are regarded as the "Future Green Engines".

It should be noted that, although nanomaterials could be more expensive, it was proved that in the automotive sector, nanomaterials' benefits outweigh their risks. Additionally, cars are high in cost by nature yet passengers are almost always willing to pay more for enhanced performance or better technology. This accelerates the rapid development and application of nano-based materials and technologies in automobiles.

Cases

MERCEDES BENZ

An invisible NANO-network that resists scratches.

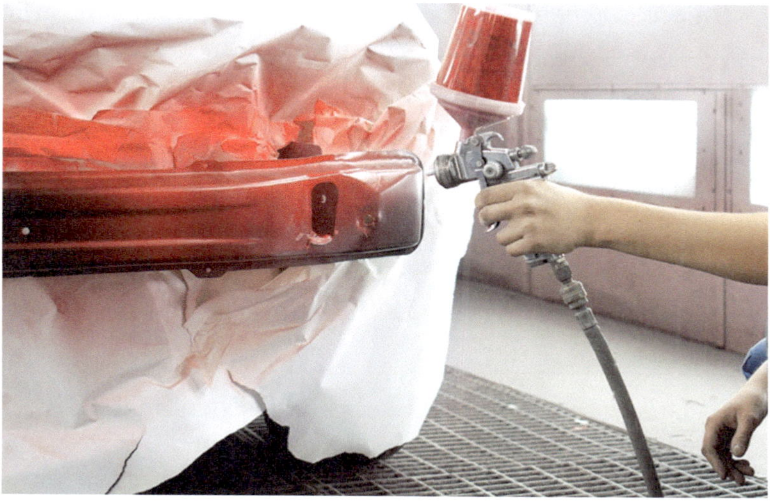

New automotive paints that meet the premium expectations of Mercedes Benz' elite customers.

Innovation

- A brilliant painting system that coats a car's body to produce a protecting shield against both scratches and chemical residues from air pollutants, moisture and washing solutions.

Nanotechnology

- Tiny silica nanoparticles that are tightly packed and cross-linked in an elastic organic binder forming a high strength and stiff network [1].

Results

- The better gloss and increased scratch resistance enhanced the long term quality and value retention of Mercedes Benz' cars to the customers.

Drivers

Vehicle numbers' growth. According to the United Nations, the worldwide vehicle market is estimated to double from 750 million in 2008 to 1 billion in 2030.

Highly competitive market. International competition drives companies to work hard and invest in new technologies that ensure the comfort, safety and eco-friendliness of future cars without compromising people's livelihood.

Rapidly developing sector. The automotive industry is an advanced technology sector that is economically important and adopts new technologies early.

Barriers

Technical conflict. Usually scratch resistant materials are not resistant to chemicals and vice versa. Thus it was incredibly challenging to maintain both of these features in the same paintwork, a limitation which had hindered the development of clear lacquers till then.

Disruptive technology. The implementation of the technology in car body paints required multi-disciplinary experts to fulfill different requirements.

Long term uncertainty. New nanotechnology developments in the automotive sector are being rapidly adopted but have uncertain issues related to durability and environmental sustainability.

Enablers

R&D. Mercedes Benz had the capabilities and resources to conduct 4 years of extensive R&D until it developed this innovation.

High standards to meet. Mercedes Benz sets stringent standards and requirements to be met in all its car features including the painting systems of its car bodies.

Extensive testing. Mercedes Benz employed a long term testing program that included more than 150 test cars that were subjected to everyday harsh conditions for several years. The results showed significant improvements.

Environmental friendliness. The base coats of painting systems meet Mercedes Benz' quality and sustainability standards because they use 80 % less organic solvents than conventional painting systems.

Impact

Threefold improvement. The cross-linked nanoscopic network system produced a threefold improvement in the scratch resistance of the paintwork compared to conventional paint.

Durable gloss. Although the technology is based on an invisible nanoscopic feature, it visibly enhanced gloss by 40 % over a long period of time.

Accreditation. Mercedes-Benz is the world's first vehicle manufacturer to offer this scratch-resistant clear lacquer. This feature is one of the features that supported Mercedes Benz' SL class to be certified by the South German Technical Inspection Authority (TÜV Süd) in 2005 with the Environmental Certificate in accordance with the ISO TR 14062 standard [2].

Dual compatibility. The New technology is compatible with both metallic and nonmetallic body finishes.

What's Next?

Manipulating the functional properties of materials at this nano-scale provides a huge potential for Mercedes Benz' scientists to enhance and modify materials in other areas of the automotive industry such as nano-structured dirt deflecting wheels.

BRIDGESTONE

Green tires that roll faster and save the planet.

Bridgestone employs nanotechnology in its tires to produce sustainable high performance tires.

Innovation	•Bridgestone utilizes a proprietary nano-polymeric compound to increase the homogeneous distribution of silica in rubber and, consequently, reduce rubber stiffness which reduces the rolling resistance.
Nanotechnology	•Nano-polymeric particles. Such nanoparticles are homogeneously distributed to ensure tighter bonds between the different materials used to make the tire.
Results	•The newly developed state of the art materials and technologies enabled Bridgestone to develop tires that offer a successful balance between environmental friendliness and safety through enhancing the wet grip performance (ease of stopping on wet roads).

Drivers

Regulatory push. The new EU tire label is forcing strict automotive standards for safety, durability and noise which is driving tire manufacturers to develop more efficient and ecologically compatible tires.

Economy. Rising fuel prices, combined with the increased pressures on personal income by the global financial crisis, have increased the need for fuel efficient safe tires.

Raw materials. The recent steep rise in natural and synthetic rubber prices has driven the need for discovering and developing cheaper sustainable alternatives.

Barriers

Safety issues. Environmental and health safety concerns regarding the potential release of nanomaterials in the environment as a result of tire abrasion or disposal were some of the main limitations.

Technical issues. Achieving good dispersion of silica in rubber for optimum chemical interaction and performance was a great challenge.

Misconception. Public and customer misconception that the low rolling resistance feature in tires could compromise the tire's durability and performance in other areas, even if it provides better fuel efficiency.

Needless investment. Open book contracts (in which buyers and transport providers agree on a fixed operational margin in response to increases in fuel prices) can wholly or partially shield operators from the increases in fuel prices and thus reduce the incentive for investing in fuel efficient tires.

Enablers

EU tire label. The new EU tire label increased public awareness about sustainability in tires, thus consumers can identify and distinguish greener, higher performance tires. Furtherly, tire manufacturers are competing more viciously to achieve higher sustainability standards in their tires.

Market size. Rubber tires are by far one of the biggest commercial markets for nanomaterials and, according to Continental, green tires take up about 1/3 of sales [3].

R&D and sustainability. Bridgestone is one of the biggest firms investing in the R&D of tires to produce innovative sustainable solutions; its newly developed tires are being fabricated with 5 % recycled rubber from old used tires.

Rate of adoption. Innovative green tires penetrate faster in premium expensive car markets where they constitute around 20 % of a mature market as price sensitivity is lower [4].

Impact

Better performance. Bridgestone was able to achieve a 12 % improvement in rolling resistance without sacrificing the high performance levels of durability, wet safety and wear resistance.

Environmental. The new tires increased added value to the consumers and the environment by reducing both fuel consumption (by up to 5 %) and carbon emissions.

Recyclability. Nanomaterials reduced or replaced heavy materials in tires resulting in lower weight tires and increased their chances of recycling.

What's Next?

Nanomaterials in tires have great potential to serve the growing demand in the car industry (expected to double by 2020) which cannot be met through current production methods and materials that rely on petrochemicals.

UMICORE'S CAR EXHAUST

Keep the air clean and make money.

Umicore has a successful whole established business unit focused on reducing and cleaning car exhaust.

Innovation

- Innovative processes that can control the fine and homogeneous distribution of nanocatalysts to obtain better catalytic effect per gram of precious metals and save valuable resources.

Nanotechnology

- Catalytic nanoparticles have a huge surface-to-volume ratio that provides the surface needed for accelerating and catalyzing slow chemical reactions.

Results

- Umicore's long and successful investment in nano-catalysis helped it utilize an excellent opportunity to expand its business in new markets.

Drivers

Regulatory push. Increasingly stringent emission regulations, not just in the European Union or North America, but also in Asia and China, put heavy pressures on the automotive industry to foster and lead technological developments in reducing car exhaust emissions.

Big chance in emerging markets. Umicore is not a main player (third position) in emerging markets (India and Russia), therefore, it has a great potential for growth in these markets through competitive technological advantage, especially that Asia is working to catch up with the stringent legislation regarding car exhaust emissions.

Company strategy. "Clean air is our business." The company is following a strategy to develop innovative solutions for cleaner air with 80 % of Umicore's R&D investments being utilized in the clean technology research domain [5].

Barriers

Shortage of resources. Although three-way catalysts efficiently reduce carbon monoxide, non-methane hydrocarbons and NOx emissions, they are not entirely sustainable products. They contain significant amounts of platinum group metals—platinum, palladium and rhodium—which mostly come from mines in Russia and South Africa, rare earth metals, most of which are mined in China and can be hard to get hold of due to Chinese export bans and zirconium oxide. All such materials are highly expensive and not abundant [6].

Regulatory upset. Despite the EU's End of Life Vehicles directive introduced in 2000 to reduce waste from passenger cars and light commercial vehicles by increasing material and component reuse, recycling and recovery, the recycling rates of the metals in these catalysts are still low (less than 35 %) [7].

Heavy consumption in one application. 82 % of the world's production of the noble metal rhodium, 51 % of the production of palladium, and 19 % of the rare earth metal cerium are presently used for the production of three-way catalysts for purifying exhaust gases of gasoline vehicles. Thus, the current consumption of such elements in car exhaust catalysis compromises their implementation in other useful areas like fuel cells, medicine and sensors [6].

Enablers

A tier company. Umicore is one of the three world leaders in automotive catalysts with a 30 % market share in the light duty market (passenger cars).

Sustainable act. Umicore is following a sustainable strategy to save valuable resources of precious metals from depletion by recovering them from the disposed automotive catalysts in one of the largest recycling facilities: Precious Metals Refining.

R&D. Around 7–10 % of the revenues of Umicore's nanocatalyst business is used to fund the company's R&D. Umicore has five R&D and test centers and one in every six of its employees works in the Research department.

NANOfutures initiative. Umicore is a member of the European NANOfutures initiative that coordinates the societal needs and impact of nanotechnology through integrating a network of reputable enterprises [5].

Impact

Cleaner exhaust. The nanocatalysts reduce carbon monoxide and nitrogen dioxide content in vehicle exhaust by 95 % [8].

New businesses. Umicore's long term expertise in developing and manipulating the nano fine distribution of catalysts gave it a strong competence in the fields of hydrogen and fuel cells, which are more environmentally friendly and are considered to be the future sources of energy.

New markets. Driven by the successful technology and the continuous growth in car markets in China (and consequently more stringent emission regulations), Umicore opened a second plant in China to encourage innovations in the automotive industry.

What's Next?

A research team in Umicore Hanau, Germany is using computer simulations to study the correlations between the chemical properties of catalytic materials and their behavior in an attempt to develop catalysts with significantly reduced amounts of precious metals and to optimize their performances to meet the stringent emission regulations [6].

MILLERS OILS

Tiny soccer balls in automotive oils!

Millers oils developed a fuel efficient formula for engine and gearbox oils based on spherical nanoparticles.

Innovation

- The nanospheres in Millers oils have large surface areas which migrate and stick to lubricated components acting as billions of ball bearings. They also fill the gaps in microscopic asperities (unevenness like microscopic peaks and valleys) on metallic surfaces [9].

Nanotechnology

- Buckyballs are nanospheres of fullerenes comprised of 60 atoms. They are surrounded by spherical carbon layers. The concentric carbon layers exfoliate under pressure creating a protective film.

Results

- In 2013 Millers Oils won the prestigious Motorsport Industry Association 'Technology and Innovation Award' in recognition of its continuous success in developing innovative products.

Drivers

Friction consumes energy. Internal friction within car engines does not only cause wearing, but also causes 15–20% of energy loss through heat buildup.

Megatrends. Fuel efficiency and CO_2 emissions became vital issues in the automotive market which drove the need for fuel efficient solutions that do not affect the durability or performance of auto parts.

Demand pull. Conventional gear oils use molybdenum sulfides that cannot withstand extreme pressures and are highly sensitive to heat as opposed to fullerenes.

Barriers

High cost. The technology is extremely costly which urged Millers Oils to introduce it primarily in the premium motorsport industry before penetrating the passenger cars market.

Risky investment. Millers oils is one of the first oil and lubricant companies to invest in nano-based oils and foresee them as the future. However, most Original Equipment Manufacturers (OEMS) prefer already proven budget products.

Consumer and customer learning. Most car users and OEMS are not aware enough of the ongoing nanotechnological innovations in engine and gear oils.

Technical conflict. Fuel efficient engines require low viscosity oils, but low viscosity oils do not adequately protect against wear.

Enablers

Synthetic oils. Nanodrive oils are fully synthetic eliminating the freight dependence on crude oils that are suffering a shortage crisis.

Motorsport paves the way. Since the launch of Nanodrive oils in January 2009, they have been used and acknowledged by a number of teams in the British Touring Car Championship as well as by many successful rally drivers such as British Junior Champion Martin McCormack and Scottish Rally Champion David Bogie.

Company expansion. Millers Oils tripled its turnover in 11 years and, in just 3 years, has have witnessed great international expansion which increased the company's dedication to investing in new innovative technologies [10].

Stop-Start vehicles. Research forecasts predict that Stop-Start vehicles will capture more than 50 % of the global market by 2022. The Stop-Start process increases attrition in the engines compared to conventional vehicles. Nano-based oils are perfectly suited for such vehicles because of the high surface area of thin films that stick and protect engine parts [11].

Impact

High performance. Miller Oils was able to produce lubricants that reduced friction. This led to decreased energy loss, increased fuel efficiency, improved durability and cost efficiency.

Spin off. The new Nanodrive oils showed great success in motorsports where high performance was especially critical. As a result, they have been implemented in the Energy Efficient Oils of Passenger Cars and achieved a 2.5 % enhanced fuel efficiency.

Award. In 2009, Millers Oils was awarded the 'Most Innovative New Product' award in the World Motorsport Symposium for its innovative product.

What's Next?

Millers oils recently invested 500,000 euros in a new Research and Development Centre in the UK to enhance its ability to develop innovative new products and to introduce these products to the market in shorter periods. In addition, it is planning to strengthen its presence in Africa and the USA [10].

GENERAL MOTORS

Stronger and lighter weight vehicle bodies.

General Motors (GM) is the first company to apply nanocomposites in cars body parts to render them lighter and stronger.

Innovation

- It is not only about using nanocomposites, but also about an innovative process that can peel apart the nanoclay fillers in a polymer. This innovative process resulted in the separation and homogeneous distribution of the nanoclay fillers within the matrix (resin).

Nanotechnology

- A nanocomposite of fine-grained nanoclay fillers suspended in a plastic resin. The fillers are evenly distributed in the resin exposing high surface area for interaction with the resin and, consequently, stronger body parts.

Results

- Since GM's successful implementation of nanocomposites in 2002 models, it has been expanding the development and the wide application of nanocomposites in the following 2003, 2004 and 2009 models. General Motors consumes around 660,000 pounds of nanocomposites per year [12].

Drivers

The expanding car market. The industry for cars grows in size every year, while the automotive sector is the primary source of carbon emissions released to the environment.

Technology push. The huge surface area of nanoclays offers exceptional improvements in the plastic properties (compared to conventional talc where only 2.5 % of the nanofillers in auto parts resulted in lighter weights) and are as stiff as those made with ten times the amounts of conventional talc.

Market pull. The automotive market is continuously demanding auto manufacturers to supply lower weight autos because of the *direct* impact on fuel efficiency and cost and the *indirect* impact on sustainability.

Barriers

Initial uncertainty. The performance was questioned in the beginning and that's why the technology was applied primarily on simple auto parts (step assist).

Cost. Nanofillers are usually more expensive than conventional ones, thus there is a need to validate their cost effectiveness.

Unstable crude oil prices. The production of nanocomposites on a commercial scale is very challenging because the prices of the polymer matrices used depend on the continuously fluctuating crude oil prices.

Enablers

Sustainability. Nanoclays are recyclable nanomaterials, with relatively lower costs compared to other nanomaterials.

Technical feasibility. The same conventional equipment can be used and no new molding tools are needed to install the new nanofillers.

R&D and collaboration. GM's R&D managed to develop the nanocomposite by collaborating with Basell North America (the world's largest producer of polypropylene resins for plastics) and Southern Clay Products (which mines and processes the ultrapure "smectite" clay additive) for more than 2 and a half years.

Impact

First entrants lead. GM is the first car manufacturer to introduce nanocomposites in car bodies in the step assist on several models.

Recognition. GM's commitment to continuously offering greener and better performing products caused it to be recognized as one of the top nanotech innovators in 2012 by the NanoBusiness Commercialization Association.

Operational. Although nanofillers are more expensive, the composite uses lower filler content than conventional fillers amount. Therefore, on a volume basis it costs less, additionally, fewer adhesives for attachment are needed.

Added value to consumer. An increased added value to the consumer because lighter weight parts lead to better engine efficiency and less gasoline consumption.

What's Next?

GM is working on controlling the application of carbon nanotubes in nanocomposites in class A vehicles.

References

1. Werner M, Kohly W, Simic M (2008) Nanotechnologies in automobiles-innovation potentials in Hesse for the automotive industry and its subcontractors. Volume 3 of the aktionslinie hessen-nanotech series of publications: the Hessian ministry of economics, transport, urban and regional development
2. (2012) Environmental certificate for the Mercedes-Benz SL-Class. Life cycle. Daimler AG, Mercedes-Benz cars, Stuttgart
3. (n.d.) http://www.bandag.eu/bandag-eu/Francais/Bienvenue-chez-Bandag!/Aperu-des-nouvelles/Aperu-des-nouvelles-Bridgestone-launches-low-rolling-resistance-Ecopia-truck-tyres.html. Accessed 7 Sep 2014
4. (2013) Progress report: case study on nanotechnology for sustainable development of tyres. DSTI/STP/NANO, 5
5. (2011) http://www.euronanoforum2011.eu/wp-content/uploads/2011/09/enf2011_openingplenary_van-sande_fin.pdf. Accessed 8 Sep 2014
6. Votsmeier M (2013) Aim of resource-efficient automotive catalysts. Presented at the Bunsentagung 2013: 112th General Assembly of the German Bunsen Society for Physical Chemistry, Karlsruhe, Germany
7. Webb F (2013) How nanotechnology could make car catalysts more sustainable. http://www.theguardian.com/sustainable-business/nanotechnology-car-catalysts-sustainable. Accessed 8 Sep 2014
8. Wolfgang L (2008) Applications of nanotechnologies in the energy sector. Volume 9 of the aktionslinie Hessen-nanotech series of publications: the Hessian ministry of economics, transport, urban and regional development
9. (n.d.) http://performanceracingoils.com/about-nanotech-ezp-6.html. Accessed 8 Sep 2014
10. (2012) Fuel oil news[Magazine]. Ashley and Dumville Publishing 35(10):1757–1057
11. Yoshida J (2013) Would you drive a stop-start vehicle in 2014? http://www.eetimes.com/author.asp?doc_id=1320510. Accessed 9 Sep 2014
12. (n.d.) http://www.nanotechproject.org/cpi/products/general-motors-r-automotive-exterior/. Accessed 11 Sep 2014

Chapter 7
Electronic Applications

The *nanotechnology era* is an expression commonly used in the electronics industry to show how nanotechnology has revolutionized this industry and is continuously contributing to its growth. Nanotechnology transformed the computing industry by continuously developing faster and smaller chips. Moreover, nanomaterials could be integrated with many electronic appliances to induce them with improved and more unique properties. In this chapter, different applications of nanotechnology in electronics are being covered. The applications range from the complex specialized use of nanomaterials in nano robots that dig for oil to the simple use of antibacterial nanocoatings in order to coat the interiors of household machines. The chapter shows the potential of a variety of nanomaterials (Nanowires, Quantum dots, nanoparticles) in a variety of products and/or processes and discusses the barriers corporations faced during the development of their products and how they could tackle these limitations.

Introduction

Nanoelectronics is a powerful successor of microelectronics encompassing nano-based building blocks as strong role players such as semiconducting, inorganic nanowires, quantum dots, nanographene and nanotubes. Nano and microelectronics are strong contributors to the global economy with nanoelectronics ranked as one of the top contributors to the nanotechnology market. That's why countries are investing hugely in the nanoelectronics sector so as not to miss the *nanotechnology era*.

It is because of nanotechnology that transistors can be fabricated in much smaller sizes resulting in smaller and much faster computers. It is because of nanotechnology that components can be fabricated into nanosizes and be integrated with radar systems and cellphones to produce highly efficient performances in telecommunications.

M. Anis et al., *Nanovate*, DOI 10.1007/978-3-319-44863-3_7

Currently there are electronic chips in computers and cell phones with regions etched up to 65 or less nm wide on the market [1]. Tier companies in the electronics industry are always under fierce competition to flood the market with newer, faster and more efficient chips. In mid-2015, IBM announced that it fabricated the first 7 nm working chip. This a great achievement in the electronics industry which would enable billions of transistors to be fabricated on chips the size of a fingernail [2].

Nanotechnology in electronics has also contributed to the lighting industry. Many nanomaterials are currently applied in Light Emitting Diodes (LED) used in displays and photovoltaics in an attempt to produce more efficient, less expensive and longer lasting lighting systems. Another emerging field in the field of nanoelectronics is Spintronics. Conventional electronics rely on developing devices using the electric charge of electrons. Spintronics rely on developing devices using both the electric charge and the spin properties of electrons. This would tremendously impact the development of stiffer, faster and smaller computer chips [1].

At these sizes, the nanoelectronics industry is facing new opportunities and new challenges because quantum effects dominate and because of the difficulties and high expenses associated with the manufacturing tools and techniques. Currently, the most common approaches in nanoelectronics are based on Top-Down methods (Nanolithography), however due to physical limitations and costly barriers, there are many R&D attempts now to fabricate them using Bottom-up approaches in which molecules and atoms self-assemble in a controlled fashion to produce the intended nanometer sizes of the electronics components.

It is also worth mentioning that there are growing trends in researching and developing electronic components that can be integrated and can interact with biologic components. Such bio-nanosensors are super-fast, super accurate and super small. They would help clinicians to better understand cellular processes at the nanoscale, diagnose diseases earlier and invent more effective treatments. The lab on a chip technology in microfluidics is under extensive research owing to the futuristic benefits expected to take place with the wide marketing and application of such high technology in diagnosis and treatment.

Cases

SONY LED TV

Better visual experience with invisible materials.

Sony TVs are employing new technologies to produce stunning visual experiences on large displays.

Innovation

- In TVs, each pixel is illuminated from behind with a white backlight and different colors are created by changing the amount of light allowed to pass through three different filters. QDs (quantum dots) eliminate the need for backlighting and enhance the purity and transparency of the produced colors.

Nanotechnology

- They use QDs (semiconductor nanocrystals) whose nanoscale features emit highly specific wavelengths.

Results

- Differentiation in the consumer electronics market is difficult but Sony managed to differentiate its new TV sets by being the first to employ QD in the TVs to enhance the picture and color quality of the TVs.

Drivers

Sustainability issues. LCDs use Cold Cathode Fluorescent Lamps (CCFL) which are not energy efficient despite being cheap as well as their mercury content that is highly toxic.

Power efficiency. The power efficiency of traditional LEDs are several 10s of percent less than QD LED because traditional LED uses many absorptive color filters to achieve the same color performance as QD that tend to absorb photons generating heat rather than useful photons as with QD.

Internal strategy. Sony is following a medium to long term environmental plan "Road to Zero" that aims at achieving zero environmental footprints through its business activities and related products. For example, CCFL that contains mercury is completely banned in Sony's new TV sets [3].

Barriers

Regulatory limitation. The main component in the applied QD is cadmium, which is strictly regulated, to the extent that it is banned in certain countries. Sony wouldn't be able to sell its new TVs except to certain limited markets.

Competitive disadvantage. QD Vision supplies Sony with QDs. However, Nanocor (main competitor of QD Vision) manufactures QDs that are cadmium free. Sony claims that cadmium is needed for a specific technical benefit in its TVs.

Limited natural resources. LEDs in general depend on critical materials that are suffering severe shortages (like indium, gallium and other rare earth metals) because of the many emerging technologies that depend on them.

Enablers

Competitive advantage. QD Vision claims that it is the only quantum dot company that is solely focused in its QD research on optical displays and lighting.

Regulatory exemption. The application of cadmium in QD Vision's optical component is exempted under the EU Restriction of Hazardous Substances Directive because there are no reliable alternatives [3].

Technical feasibility. The QD model used in the TVs can be incorporated into old LCDs with certain modifications rather than dramatically changing the manufacturing process.

Complementary products. Sony markets complementary products that are compatible with the QD enhanced TV display like digital cameras and Blu-ray disc players.

Impact

Elegant product. The new TV sets are slimmer and have wider color gamut that provides more appealing, natural and vivid viewing experience.

First entrants. Sony HD Bravia are the first TV sets to use QD technology in TV sets.

Operational. QD technology are found to be flexible and compatible with all screen sizes, in addition to longer lifetimes compared to current OLEDs (organic based LED).

New markets. The new display technology showed great success in the sales of Sony HD Bravia that it was implemented in Sony's HD smartphones.

What's Next?

Sony is exploring cadmium-free technology as a more abundant alternative to expand its products' marketing in more and new markets especially those where cadmium use is banned.

LG ALL IN ONE PC

Touch screens are getting bigger than ever.

LG partnered with Cambrios to be the first to apply its newly developed nanomaterial-based touch panel in a new model LG All-in-One PC.

Innovation

- LG designed extremely thin silver nanowires that could substitute Indium Tin Oxide (ITO) and produce a new generation of more efficient and cheap electronics.

Nanotechnology

- The company used silver nanowires that randomly cross each other producing a transparent surface and at the same is highly conductive. Such wires have high aspect ratio; tens of nanometers wide and tens of micrometers long.

Results

- The developed LG's touch panels are lighter in weight, thinner, stronger and cheaper than the glass (ITO) based touch panels saving the consumers around 30–50 dollars on bigger displays [4].

Drivers

Market size. The market for touch screens is booming in a way that exceeds all kinds of displays in the market and is expected to double yearly until 2016.

Old is not gold. The currently used indium tin oxide materials in touch screens is not helping with large screens because if the touch signal has to travel larger distances, it becomes weaker and the screen doesn't respond effectively.

Unmet future demand. A large portion of the touch screen market is reliant on ITO. Knowing that indium is a rare earth metal and that the touch screen market is growing rapidly, the need for more abundant alternatives is becoming very critical.

Barriers

Up-scaling. It is particularly challenging to transform successful applications of nanomaterials from the laboratory scales to commercial level scales.

Early stages of adoption. It is too early for touch functionalities to completely displace the keyboard and mouse in devices with larger screens and many specialized applications are needed to complement their application on the new devices.

Smartphones' markets: not yet. Although a 5-in. smartphone screen using silver nanowires would save $1 an inch, this would not make huge cost savings in the ultimate retail price. That's why large screens are currently the primary target [5].

Enablers

Newer bigger users. The launch of Windows eight facilitated the introduction of touch functionalities to devices with larger screens such as the All-in-one PC, laptops, and Ultrabooks.

Partnership. New entrants in an existing market must offer an advantage. Thus when LG decided to penetrate the touch module manufacturing market, it partnered with Cambrios to make use of Cambrios' proprietary nanomaterial.

Conventional manufacturing process. Coating substrates with the newly developed silver nanowires solutions can be accomplished by standard roll to roll printing techniques, which are cost effective and fast compared to the deposition techniques used with ITO [4].

Impact

Operational. The nanowires enabled LG to achieve super high conductivities on the large screen monitors which was unachievable with ITO.

Lead in growing markets. The new technology placed LG in a leading position in the rapidly expanding market that demands cost effective touch functionalities on large screen panels.

Competitive advantage. LG is not a new entrant with its devices in the touch screen market, thus nanotech based touch screens provided LG with a unique competitive feature in their newly introduced devices.

What's Next?

The transparency and flexibility of the new nanomaterial based displays show great potential in the next era of flexible, curved, rollable and 3D featured displays.

ARAMCO SAUDI NANOBOTS

Nano-robots can dig for oil and gas.

Saudi Aramco for oil and gas industry pioneered the invention of Resbots: nanobots that can travel through the nanopores of reservoirs to explore the oil trapped there.

Innovation

- Miniaturization of electronic machines to go through the narrowest channels which are otherwise unreachable by conventional machines.

Nanotechnology

- Resbots are nanoelectronic devices (1/1000th the size of human hair) that can sense, explore and provide clear images of oil reservoirs.

Results

- It is not typical for national companies like Saudi Aramco to invest hugely on R&D but its continuous investigation of novel technologies is putting it on the right track to increase its average oil recovery to 70 % in the coming few years (double the world's average oil recovery) [6].

Drivers

Unrecovered oil. Based on existing technologies, it is estimated that around 50 % of the discovered oil fields in Saudi Arabia are still unrecovered due to limited technologies [7].

No room for errors. In an era where oil demand exceeds supply, oil companies have to work on developing methods of exploring their reservoirs to mitigate the risks, costs and time loss associated with dry well drilling.

Demand pull. Reservoirs are not open tanks; they are solid porous rocks with oil trapped in these pores. The Exploration and Petroleum Engineering Center—Advanced Research Center (EXPEC ARC—Aramco) conducted several analyses that revealed that most of the pore throats are greater than or within 500 nm, which drives the need for miniaturization techniques.

Barriers

Semi radical innovation. Most of the research and work conducted on nanobots was aimed at targeting medical and treatment applications; little data at that time was available about applying nanobots in exploration of oil reservoirs.

Technical. Saudi Aramco researchers faced great challenges to control the stability of their developed nanoparticles in the high salinity and temperature conditions of the Saudi Arabian deserts [8].

Mass production. Miniaturization to the nanoscale costs too much, especially when mass productions of tons of these nano-agents are needed for field work at reasonable prices [9].

Enablers

Open innovation. 'Nanoscale Revolutions to Mega-scale Challenges?' is the title of a workshop organized by EXPEC ARC where it gathered researchers, analysts and engineers from major oil companies, well known academic institutions and even from companies outside the oil industry to discuss the future of its proprietary nanobots, potential challenges and solutions.

Successful field trials. Since the first introduction of Resbots by Saudi Aramco in 2007, the company conducted several field testing and trials to manipulate and optimize the efficiency of these nanobots. One test analyzed 850 core plugs in the Ghawar field to determine the nanoparticles' optimum size range [10].

Strategic goal. In the last 10 years, Saudi Aramco has increased its annual capital budget from 4 billion dollars to 40 billion dollars, with investments in-house R&D and technology in an attempt to become one of the world's pioneers in energy generation by 2020 [6].

Impact

Award. Research and innovation of Resbots earned Saudi Aramco the prestigious World Oil Award in 2008 and since then it has been leading the industry developing and optimizing these nano agents [8].

Explore the unexplored. The nanobots enabled the company to explore and penetrate regions in reservoirs that were never explored by previous technologies.

Investment economics. The improved process efficiency associated with the use of Resbots could enhance the recovery of oil and gas and increase cost savings.

What's Next?

Saudi Aramco is investigating two types of the next generation Resbots: the active one that can sense the reservoir environment conditions and the reactive one which can deliver chemicals to intervene with these conditions [6].

INVISAGE'S HIGH RESOLUTION IMAGE SENSORS: QUANTUMFILM™

High resolution ultra-thin Cameras.

InVisage made new Image Sensors that allow a higher zoom, resolution and ultra-thin cameras.

Innovation

- A photosensitive layer designed instead of the conventional complementary metal-oxide semiconductor (CMOS) image sensors to increase sensitivity to light and electric storage. This layer enables the camera to detect more photons, store more electrons and reproduce colors more accurately—all with a thinner camera module.

Nanotechnology

- A colloid of very small semiconductor nanoparticles called quantum dots. Quantum dots have electrical properties that are between the bulk form and molecule form. Their electronic and optical properties depend on their size and shape [11].

Results

- The use of Quantum Dots produced a very efficient image sensing layer that has superior properties and efficiency over the conventional CMOS silicon semiconductor with a much stronger absorption efficiency and high resolution capability.

Drivers

Cameras are getting smaller. With the increasing use of mobile phones and other small handheld devices that have cameras in them, this has led to the increasing need to make cameras smaller while maintaining or even improving their image taking capabilities.

The need for high resolution, high quality images. There has been an increasing need for high resolution high quality images due to the increased awareness and demand of customers.

The increasing frequency of taking pictures. With the increasing use of social media like Instagram and Facebook, people want to capture images of themselves at all times i.e., in bright and in even very dark lighting conditions, which necessitates an image sensor capable of adapting to all lighting conditions.

Barriers

The technology. The technology used is very complex and would require facilities and experience to be implemented.

Instability of Quantum Dots. Quantum dots can be oxidized if exposed to air. The protection InVisage used, by placing them underneath a color filter and electrode layer though, prevents this from happening. Moreover, during manufacturing, the use of ligands to protect them has been employed [12].

Enablers

Research. InVisage has been doing research in this technology since 2006, giving it the experience necessary to manufacture these products.

Collaborators and Investors. The company has a number of collaborators and investors like the University of Canada, Nokia Growth Partners and Intel Capital.

Several Offices. It has several offices: a headquarters in Menlo Park, California, a small design site in New Hampshire, Canada and a growing office in Hsinchu Science and Industrial Park, Taiwan.

Impact

High Quality Products. This technology has enabled InVisage to provide high efficiency products to mobile devices and even to professional cameras.

Higher Absorption Efficiency. The QuantumFilm™ has a much stronger absorption efficiency than conventional silicon films. The reason for this is that silicon is a weak absorber of red light; the light bounces around the silicon atoms before transferring the energy to it making absorption less efficient. In QuantumFilm™ on the other hand, light doesn't bounce and passes its energy directly to the electrons enabling a fast and efficient transfer of energy from the light to the quantum dots to create an electric signal. In addition, InVisage has made the QuantumFilm™ layer at the top surface, unlike the silicon in conventional sensors where they are at lower layers, as they are used to do both light sensing and electronic read-out functions. Thus, the QuantumFilm™ is in a place to absorb more light than the silicon.

No bright light saturation. In conventional image sensors, saturation takes place in bright light where the light photons cause an overflow of electrons in the sensors storage area causing loss of picture detail. In the case of QuantumFilm™, the light sensing is done by the film in one plane and the processing of the signal takes place in a different plane resulting in a greater storage space for electrons to pass and preventing saturation from occurring. As a result, the dynamic range is greatly enhanced producing less saturation in bright conditions and greater detail in dark ones.

Much thinner cameras. With silicon-based sensors, there is a limit as to the height of the camera lens—when it becomes shorter than a certain value color error—blurring and washed out images, especially at the edges of the image sensor. This doesn't happen with the QuantumFilm™ allowing for thinner cameras and steeper angle focusing.

Pictures of moving objects. An additional advantage of the QuantumFilm is that it can take pictures of moving objects without distortion of the image unlike the conventional CMOS sensors.

Higher zoom capability. QuantumFilm™ also allows more zooming than the conventional image sensor.

Infrared sensing. QuantumFilm™ senses longer wavelengths and infrared light that are invisible to the human eye.

What's Next?

With InVisage's experience, it can use the quantum dots films for other applications like in displays of mobiles and other portable devices. According to the recent studies, quantum dots improve power efficiency by 50–100 % while maintaining the same brightness of images [13].

SAMSUNG'S WASHING MACHINES

Washing machines that kill bacteria.

Samsung invented new washing machines with antibacterial activity. The production of silver ions during washing imparted an antibacterial activity on the washing machine

Innovation

- New washing machines with antibacterial activities. Such washing machines are especially important in case of using cold water to reduce energy and water consumption. This is because, in these machines, the cleaning ability is reduced and thus leads to biofilm formation. Biofilm represents a community of bacteria that adheres to inert or living matter.

Nanotechnology

- Silver nanoparticles are produced during washing via electrolysis. Silver nanoparticles penetrate deep into clothes sanitizing them even with cold water. Moreover, silver nanoparticles help eradicate biofilm.

Results

- The new technology allows more than 100 quadrillion silver ions to deeply penetrate the fabrics sanitizing them and thus eliminating the need for hot water and bleach. Moreover a wide range of products applying the same technology have been developed including air conditioners, refrigerators and vacuum cleaners.

Drivers

Nano silver trend. The nanotechnology market is increasing tremendously, and with it, the fields of silver nanoparticles research and products are swiftly growing [14].

Strong market competition. Daewoo is one of the strong competitors of Samsung and with Daewoo's lead and success in nano-enabled products, Samsung had to follow the same path so as not to lose part of its customers to Daewoo.

Barriers

Nano-silver toxicity. The debate about silver nanoparticles' safety is a major barrier due to conflicting studies concerning its safety. It requires appropriate extensive studies to ensure the safety of the newly developed product [15].

Learning process. An intensive learning process is required to increase the public acceptance of the developed product by advertising its improved properties and complete safety.

Environmental hazards. The amount of silver washed into waterways has to be precisely detected so as not to impose an environmental risk. It was calculated to be about 12.2 Kg of silver that would be washed in waterways in Germany alone. Such a concentration, according to non-governmental organizations, needs to be considered as there are other silver sources from textiles and other products. Overall, the whole percentage of silver may be risky for the environment.

Public pressure; withdrawal and release. Public concern about the fate and safety of the silver nanoparticles resulted in temporary withdrawal of Samsung's nanosilver washing machines from sale in Sweden, however, they were later reintroduced into the market. Nevertheless, public pressures in other countries as Germany and Austria continues [16].

Enablers

Strong R&D. Samsung's strong R&D allowed it to analyze Daewoo's action and apply it to its products as fast as possible so as not to lose any of its market shares.

Daewoo's trial and success. With Daewoo being the first to introduce this technology, Daewoo took the risk of being in the market lead and at the same time introduced the technology and its success to its competitors.

Lean regulatory rules. The regulatory gaps present within the available guidelines allowed the introduction of nano-based products in the market [17].

Impact

Internal. The introduction of nanosilver technology extended to a wide range of Samsung's products including washing machines, refrigerators, vacuum cleaners and air conditioners, thus allowing it to be a strong competitor in an increasingly competitive market.

Market. The nanosilver-based technology was first introduced by Daewoo, yet with the efforts of Samsung, the market moved from being a monopoly for Daewoo to a competitive market with strong players like Samsung, LG and Daewoo.

Regulatory bodies. Due to public pressures from different NGO's, the US EPA will attempt now to regulate nano-based products that contain nanosilver as pesticides.

What's Next?

Samsung added silver nanoparticles to its washing machines to benefit from their antimicrobial properties. Such technology could be extended to any product or application in different fields that could benefit from being antibacterial. Examples include:

- Antimicrobial surfaces as in hospitals, public places, food and beverage factories
- Refrigerator boxes and kitchen ware
- Baby toys, baby car seats and kitchen baby chairs
- Hospital bed dressings
- Clothes as socks and shoes

DAEWOO REFRIGERATORS

Refrigerators that kill bacteria.

Daewoo produced antibacterial refrigerators through the inclusion of silver nanoparticles in the body of the refrigerators. Thus, sanitized refrigerators with no foul odor.

Innovation

- Refrigerators with antibacterial activity, thus no foul odors are developed and better preservation of food is achieved.

Nanotechnology

- Silver nanoparticles, included in the plastic resin of the body of the refrigerator. Such silver nanoparticles prevent the growth of the bacteria, and thus suppress the foul odors and enhance the preservation of food.

Results

- Complete line of products applying the same technology including washing machines, air conditioners and vacuum cleaners.

Drivers

Nano silver trend. The nanotechnology market is showing a tremendous increase and is forecast to grow to 1 trillion dollars by 2015. Within such a growing market, more than 260 nano-silver products have been identified in different fields [14].

Daewoo's internal strategy. Daewoo is committed to continue providing health oriented products and also environmentally friendly ones. Thus, the nano-silver technology models just fulfill the company's internal strategy.

Market lead. The continuous research and development has made Daewoo the first company employing polypropylene technology in the manufacture of rustproof washing machines. Being used to leading the market, utilizing nano-silver will also make Daewoo the first company applying nano-silver in its commercial electronic production.

Barriers

Safety. The fear of migration of silver nanoparticles from food, contact material, to food coupled with the lack of definite data on the safety of silver nanoparticles stand as a barrier against applying silver nanoparticles in different products [15].

Public acceptance. Public Acceptance is a main barrier against the development of new nano-products and processes within the food sector.

Required expertise and infrastructure. The development of the new silver poly technology required research and development as well as certain expertise in the field in addition to the infrastructure for the implementation of the new technology on a commercial scale.

Enablers

Strong R&D. Through strong R&D, Daewoo proved that the addition of silver nanoparticles via injection molding is more effective than addition of silver via coating. Despite coating being less expensive, it was liable to wearing off. Additionally, the disinfectant properties in the injection molding proved to be stronger.

Enabled technology. The injection molding technology that is used for applying silver does not need completely new infrastructure but only mixers for chemical synthesis of the silver nanoparticles and injections for applying silver nanoparticles during molding.

Lean regulatory rules. Regulatory gaps concerning the new nanomaterials and technologies has permitted the introduction of nano-based products in the market [17].

Impact

Internal. The developed nanosilver poly technology by Daewoo allowed it to extend the technology to its air conditioners and washing machines.

Industry. Daewoo's invention has made its competitors (Samsung and LG) in the market adopt the technology and thus increased competition.

Competitive advantage; market lead. Daewoo, being the first to introduce the technology, allowed itself to have the market lead in the field, especially with the adoption of the nanosilver technology by its competitors.

What's Next?

Daewoo used the antimicrobial properties of silver nanoparticles for imparting disinfectant properties to its refrigerators. Such a technology could be extended to other electric appliances like air conditioners and washing machines (already accomplished). Moreover, it could be extended to any surface or container that can benefit from having antimicrobial properties in different fields such as baby toys, food factories' floors, benches, hospital walls and bed dressings.

References

1. Allsopp M (2007) Nanotechnologies and nanomaterials in electrical and electronic goods: a review of uses and health concerns. GRL-TN-09-2007
2. (n.d.) http://www.techrepublic.com/article/how-ibms-new-7nm-chip-busts-moores-law-changes-future-of-computing/. Accessed 20 July 2015
3. (n.d.) http://www.sony.net/SonyInfo/csr_report/environment/chemical/products/index1-1.html. Accessed 25 July 2015
4. (n.d.) http://www.businesswire.com/news/home/20121030005755/en/Cambrios-LG-Electronics-Announce-Worlds-23%E2%80%9D-Touch#.VhacHOyqqkp. Accessed 26 July 2015
5. Branscombe M (2013) The future of touchscreens revealed: bigger, cheaper, bendier. Techradar.pro.http://www.techradar.com/news/mobile-computing/tablets/the-future-of-touchscreens-revealed-bigger-cheaper-bendier-1168898/2. Accessed 28 July 2015
6. (2014) Pushing boundaries, NOCs advance global R&D. (n.d.) http://www.aramcoservices.com/AramcoServicesCompany/media/Hero-Areas/0114HEP-COVER-JAN.pdf. Accessed 28 July 2015
7. Alnuaim S (2013) Global oil reserves. Saudi Gazette. http://www.saudigazette.com.sa/index.cfm?method=home.regcon&contentid=20130504164199. Accessed 30 July 2015
8. (n.d.) http://www.aramcoexpats.com/articles/2010/07/nanobot-trial-a-winner. Accessed 30 July 2015
9. (2012) Illuminating Saudi reservoirs. JPT 64:4
10. (2008) Nanorobots: Aramco's technology getting small in a big way. BI-ME. http://www.bi-me.com/main.php?id=18067&t=1. Accessed 30 July 2015
11. Murray CB, Kagan CR, Bawendi MG (2000) Synthesis and characterization of monodisperse nanocrystals and close-packed nanocrystal assemblies. Annu Rev Mater Res 30(1):545–610
12. Chuang CM, Brown PR, Bulović V, Bawendi MG (2014) Improved performance and stability in quantum dot solar cells through band alignment engineering. Nat Mater 13(8):796–801
13. Chen J, Hartlove J, Hardev V, Yurek J, Lee E, Gensler S (2014) P-119: high efficiency LCDs using quantum dot enhancement films. SID Sym Dig Tech Pap 45(1):1428–1430
14. Kimbrell GA (2008) Citizen petition for rulemaking to the United States environmental protection agency. International center for technology assessment. http://www.icta.org/nanoaction/doc/CTA_nanosilver%20petition__final_5_1_08.pdf. Accessed 9 Dec 2008

15. Chaudhry Q, Scotter M, Blackburn J, Ross B, Boxall A, Castle L, Aitken R, Watkins R (2008) Applications and implications of nanotechnologies for the food sector. Food Addit Contam 25:241–258
16. (n.d.) Concerns about nanotechnology washing machine. http://www.nanowerk.com/news/newsid=1037.php. Accessed 24 July 2015
17. Senjen R, Illuminato I (2009) Nano and biocidal silver: extreme germ killers present a growing threat to public health. Report of Friends of the Earth. http://alteredstates.net/barry/newsletter408/FoE_nanosilver_report_textonly.pdf. Accessed 15 Aug 2015

Chapter 8
Printing Applications

It is astounding that an elaborate and new science such as nanotechnology can be involved in printing. The benefits of nanotechniques and nanomaterials have enabled major advancements in the printing capabilities of nowadays printers. It has definitely allowed for high resolution images that are of a wider color range than before but that the extent of it. How nanotechnology opened new doors in printing will be discussed in this section.

Introduction

If we look at the history of printing we will find that the paper, ink and polymer films were all made from dispersions of particles of one material in another. Long ago inks resembled very much the colloidal suspensions of nanoparticles nowadays but in much larger sizes. With the advent of nanotechnology, it seemed reasonable to make smaller sized ink particles in the nano-range to improve resolution of the images. It was found that not only did it improve the resolution, but also improved the color palette and extended it to include a wider range of colors and color combinations. In addition to ordinary printing, three dimensional printing is something that has been discovered recently and is gaining wide popularity in a wide range of applications. It is sometimes called additive manufacturing and involves the making of a three dimensional object by depositing layers over layers of material. Additive manufacturing is applicable in a wide range of fields including apparel, vehicles, construction, electronics, mechanical parts and medicine including tissue engineering and making artificial limbs.

M. Anis et al., *Nanovate*, DOI 10.1007/978-3-319-44863-3_8

Cases

LANDA

High resolution nanoink.

Landa has designed a new form of ink that uses nanoparticles to improve the resolution of printed images and increase the possible color palette while decreasing the amount of ink used.

Innovation

- Landa designed a printing technology that produces high resolution printing up to the nanoscale. This technology enables the extremely efficient and precise low ink lay-down to produce exceedingly round dots with super-sharp edges and high optical uniformity and consistent density. This technology is efficient with glossy or matte paper and on any other substrate.

Nanotechnology

- The company's innovation is based on the reduction in the size of pigments to a nanosize scale which makes them more powerful colorants.

Results

- When the pigments were reduced to the nanoscale, better resolutions were obtained due to more precise round edges. They also achieve a light dynamic range wider than any other printing process. They cover more Pantone colors than conventional printing and practically eliminate the need for spot colors, off press color mixing and on-press ink replacements. They also enable printing on any sort of substrate without any priming or pretreatment.

Drivers

Growing printing market. The printing market is a growing market: it is forecast that the digital print market will reach 187.7 billion dollars in 2018 at an annual growth rate of 7.4 %.

High resolution, high quality and high speed. With the advent of the immensely high resolution digital cameras and other photo taking devices, it has become necessary that when printing such images on paper, they still maintain their resolution and color.

The need for environmentally friendly materials. Landa uses water as solvent for its nano ink which reduces the release of organic solvents to the environment. The reduction of pigments to nanosize also allows reduced use of such materials to produce the same effect, which reduces the release of such pigments to the environment. A third point is that Landa provides its ink in concentrated forms to reduce the size of its recyclable containers and hence reduces transportation costs.

Barriers

Instability of the nanopigments. Nanomaterials are usually unstable due to their distinctly small sizes and high surface free energy which can cause the clumping of pigments, however, the presence of additives in the Landa ink that is produced by their laboratory prevents this from happening.

Enablers

Rich Intellectual Property Base. Landa has over 800 patents granted to its owner which gives the company the upper hand because it has significant know-how concerning the technology.

Long Work Experience. The owner of Landa (Benny Landa) established his first printing company in 1977 that became massively popular in 1993 but was then acquired by Hewlett-Packard. He then established the Landa Nanoprinting Company in 2002. Due to his lifelong experience, he was able to develop applicable new printing technologies based on the nanoscience.

Granted Awards. The Landa Company and its owner earned many awards from Israel and the USA for being so impactful on the printing industry through developing and implementing new successful technologies.

Wide range of market segments. The capabilities of the nanoink has enabled Landa to target several applications including publishing on ordinary paper, flexible printing, carton printing, general commercial printing, advertising, direct mail, point-of-sale/point-of-purchase, catalogs, transactional production, retail floor stands and counters, end caps, power wings or sidekicks displays as well as exhibition and trade show materials.

Impact

Landa has become one of the most popular and strong companies in printing (providing printing presses and ink for high resolution nanoprinting) due to the foresight of its owner in the use of nanotechnology to enhance his products.

What's Next?

The next step for Landa technologies may be to take his innovations further and develop three dimensional printing technologies as well. Three dimensional printing is the next big thing in the manufacturing of products.

BARE CONDUCTIVE ELECTRIC PAINT

Electric circuits made just from paint!

Bare Conductive from the UK was capable of making paints that conduct electricity.

Innovation

- Bare Conductive has developed a paint that conducts electricity that can be used to draw functioning circuits on paper, cardboard, fabric, glass or other materials for education as well as entertainment purposes. They also made it safe enough to be used to draw on the skin surface.

Nanotechnology

- They used small conductive nanoparticles of carbon, e.g. flakes of graphite, dispersed in a gum binder to make the paint flexible when it dries.

Results

- Bare Conductive was able to design a portfolio of products that can be used by kids, teachers, hobbyists for fun and education. Their products include paints, touch board, batteries, LEDs, sensors and other accessories to make the circuits work. Their products are sold individually or as complete idea project kits like a flashing card set or a glowing village set.

Drivers

Need for more creative greeting card ideas. Greeting cards is a large business that has a large customer base but is a mature market that is highly competitive. There are always new ideas being brought up every day to make a certain company more competitive and enable it to increase its market share.

Need for more educational toys. Nowadays there has been an increasing trend to make toys educational to develop a child's social, intellectual, physical and/or emotional development. Bare Conductive helps children of all ages learn fundamentals of electronics in a fun way including polarity, resistance, voltage or circuitry.

Science advancement. The advancement of science has driven people to invent new ideas and products. Nanotechnology has greatly advanced especially in the conductive ink and paint area. Several paints have developed using silver, copper and conductive polymer nanoparticles.

Barriers

Safety. The conductive paint made by Bare Conductive is designed for children to play with and for drawing on skin; its safety is a tremendously important thing. The safety of nanomaterials is still under debate and there is always fear that it may be the cause for diseases such as cancer.

Learning. Though science has advanced in the conductive paint area and its use in circuits, smart clothing and many other applications, the idea is not popular yet among common people due to few commercialized products.

Small startup. Bare Conductive was a small unpopular startup company set up by four designers in the Royal College of Art. They were young and had no funding.

Enablers

Diversity. The company is formed of a group of young people who have a variety of experiences in engineering as well as art design which makes them capable of designing several great idea products to be sold.

Market acceptance. Due to their fun designs that reached out to children and their reasonable prices, their products have been used by a lot of teachers and parents. This led them to be accepted around the world, selling their products in Asia, America, Australia and Europe.

Finding funding. In 2010, Bare Conductive won 100,000 British pounds from Innovate UK (formerly known as the Technology Strategy Board) in the Launchpad competition designed to foster disruptive solutions among start-up companies.

Workshops. Bare Conductive conducted many workshops and tutorials for children and old people to introduce them to the company's technology and make it more popular. It also strengthened its viral presence on social media to connect with its future or present customers.

Impact

Due to the fun and educational value of its products, Bare Conductive was able to sell many products and have more than 80 distributors around the world including the big distributer RadioShack.

What's Next?

Bare Conductive is considering developing and researching more accessories like touch boards and sensors that are compatible with its kits.

DNA TECHNOLOGIES' DNA MATRIX™ HOLOGRAMS

Anti-counterfeit label made from invisible DNA!

DNA Technologies was able to devise an invisible label made from a synthetic DNA sequence that can be detected by special equipment.

Innovation

- Security Solutions Group designed a low cost brand protection and currency anti-counterfeit solution.

Nanotechnology

- The company used strands of synthetic DNA molecules as invisible markers in the visible label's ink or holomark and devised specific handheld scanners for them.

Results

- An efficient counterfeit detection method that cannot be replicated or detected by ordinary methods and can only be detected by the company's specialized scanner.

Drivers

Increased counterfeiting. Increased forgery and fraud has driven most governments and countries to apply protective measures to their currencies through holograms, invisible inks and labels. In addition to currency counterfeiting, there is brand and goods fraudulence. With the advent of brands and their increasingly high prices, cheaper inauthentic copies are being made and sold.

Economic losses. The presence of inauthentic copies threatens the original brands as people tend to buy the cheaper alternatives rather than the authentic ones, affecting the profits of the original brand's company.

Need for security system that cannot be replicated. Since long ago anti-counterfeit measures have been used to protect currencies. However, as these measures became more widely used, they became known and easily imitated. This necessitates the presence of methods that cannot be easily imitated and that can only be read using a special device.

Increased consumer awareness. Ever since the consumer act in the early sixties, people have become more aware of their rights and what they should receive in terms of quality and price. There has also been an increase in regulatory bodies that protect the consumer.

Barriers

Conventional methods of marking. Conventional methods of marking are much cheaper than the synthetic DNA strands technology although that makes the conventional methods more easily penetrated.

Research. The ability to chemically modify DNA and isolate it while keeping it intact then reapply it requires exceptional research and technologies to avoid its destruction due to its fragile nature.

Instability. DNA is exceedingly fragile and can easily decompose if subjected to high temperatures, moisture or chemicals; however, the synthesized DNA by DNA Technologies has been fabricated in a way to protect against such degradations.

Enablers

Experience. DNA Technologies has built a lot of experience since it started in 1993. It has been able to acquire a number of patents and make effective products in the market.

Product tracking. DNA Technologies has been able to develop a variety of DNA markers and special readers which has enabled the precise tracking of such products.

Customization. The company was able to customize its technology to be applied in a variety of applications such as apparel, currencies, beverages, pharmaceuticals and tobacco.

Impact

Due to the quality and importance of its anti-counterfeiting marking system, it was used to protect big brands as those of Thomas Kinkade, Major League Baseball, Hanna-Barbera and the 2000 Sydney Olympics. It also was applied in many fields including the pharmaceutical field protecting against drug counterfeiting.

What's Next?

DNA Technologies should start selling in different countries because the importance of brand protection and product tracking has become well known around the world.

Chapter 9
Textile Applications

The textiles industry serves many different segments in a diverse market. Because the market is mainly segmented to personal textiles (normal clothing including sports textiles), technical textiles (the textiles used by labor, firemen and military) and the fabrics used in households and automobiles. This chapter shows how nanomaterials could impart unique and innovative features to the fabrics used in different segments, for example self-cleaning, ultraviolet blockade, wrinkle resistant or even flame resistant. In addition, nanotechnology facilitated the integration of wearable electronics with textiles. Such wearable electronics could help patients monitor their medical conditions regularly or could enable sportsmen to monitor and improve their performances. Nanotechnology currently enjoys an excellent opportunity to penetrate the textiles market strongly because the high prices associated with the development and manufacture of a single nanofeature could be compensated by applying such feature in different segments in a massive market at the same time.

Introduction

Textiles are one of the most important segments in consumer goods. Nanotechnology has added new functionalities to the different markets of textiles, including automotive, household and automotive textiles. Nanomaterials are currently integrated within textiles to render them UV-protective, stain resistant, fire resistant, antimicrobial and self-cleaning. In the past few years, the textiles industry had reached a state of dormancy with low cost producers in less developed countries taking the lead in manufacturing. Since the evolution of nanomaterials, western countries began to gain new competencies through the manufacture and commercialization of high tech textiles.

M. Anis et al., *Nanovate*, DOI 10.1007/978-3-319-44863-3_9

Nanotechnology imparted extraordinary properties on textiles that increased a consumer's demand and willingness to pay for such high-tech textiles and fabrics. Nanomaterials can be integrated with textiles through one of three means:

1. Within the fiber core by dispersing the nanomaterials within the fiber matrix.
2. As a coating onto the fabrics (most common).
3. Electrospinning of certain polymers into fibers in the nanometer range which produced fabrics with higher stiffness and better performances according to the type of polymer used.

The textiles industry is considered one of the fastest growing markets for the adoption of nanotech based products. This is due to the fact that consumers now are living rapidly changing lifestyles, and so tend to seek more comfortable and luxurious textiles. Another reason is that the textiles industry provides optimum substrates for the incorporation of nanomaterials owing to their normally available large surface area.

An important segment of the textiles market is the technical textiles; these are textiles for military, medical, sports or labor applications. This segment is swiftly growing because, with adopting nanomaterials, the performances of these modified textiles outweigh their costs. Another growing trend is the integration of electronic sensors into the clothing, especially in medical textiles where, for example, sensors can be integrated with clothing for cardiovascular patients to continuously monitor their heart status. Such sensors could also be used in athletes' clothes to continuously monitor and improve their physical performance.

Cases

ADIDAS LONE STAR SPIKE

Beijing 2008 Olympics goes nano!

Adidas is the first company to engineer special running shoes using nanotechnolog`y. They were designed specially for the Texan sprinter Jeremy Wariner.

Innovation

- Adidas' Lone Star Spike are the first running shoes engineered with carbon nanotubes in a special design to provide an optimum compression/penetration ratio to aid the sprinter to cope with the elasticity of the track surface.

Nanotechnology

- Carbon Nanotubes (CNT) reinforced plates and spikes. CNT have a high aspect ratio producing high strength at lower weights.

Results

- The nano engineered running shoes added to Jeremy Wariner's performance and physical confidence in the Olympics. He won the silver medal in the men's 400 m race in the Olympic competitions.

Drivers

Technology push. Nanotechnology has a strong potential in the sports wear markets because performance outweighs costs; there is a cost/benefit trade off which is accepted by consumers.

Market pull. There is a growing market trend for improving the sports' wear which directly affects sports wearers' comfort and performance driving the need for the development of innovative fibers and fabrics.

Market size. In 2008, 19 % of Europe consumers' expenditure was on sports clothing and footwear [1], and in UK sportswear is around 10 % of the total clothing market [2].

Technology forecast. Nanotechnology in running shoes made a performance difference of 1–2 % for the top 25 athletes and its implementation in fabrics is expected to grow.

Barriers

Public fear. Public fear of the impact of nano-enabled sportswear, in addition to the lack of regulatory standards that assess the safety of these products, are some of the major barriers.

Upscaling. Scaling up processes of nano-enhanced products pose technical and cost difficulties towards wide commercialization, especially that CNT are expensive nanomaterials.

Technical challenge. Designing high performance running shoes is a challenge in itself because the shoes had to be soft enough to absorb impacts yet hard enough to not flatten in a short time.

Enablers

A tier company. Adidas is one of the limited multinational companies that leads the sports textiles industry and is able to bring innovations to global markets.

Value chain. Applying nanotechnology to fabrics' finishing has a bigger profit margin and lower risks of losses compared to developing entirely new nanofibers. That's why companies like Adidas are following this strategy in developing nano-functionalized sportswear.

Deep analyses. Adidas' engineers and designers used high speed video and pressure mapping to study the running behavior of the 400 m Texan runner Jeremy Wariner for more than 2 years to custom design specialized running shoes for his running style.

Impact

Intrinsic features. The developed innovative plate was about one third the thickness of Wariner's previous spike plates and weighed 50 % less. The mechanical and chemical bonds within the nanotubes inside the spike plate increased its structural integrity and durability.

Unique performance. Knowing that long sprint races are lost in corners. Adidas lone star spike aided Wariner to push better in turns while maintaining his stability. Thus, he was able to win the silver medal in the men's 400 m race in the Olympic competitions.

Fast market adoption. Shoes worn by elite athletes allows for the ease of commercialization and rapid market adoption of such nano engineered products by average consumers.

What's Next?

Although carbon nanotubes are primarily used in aerospace and automotive industries, the Lone Star Spike shoes lead the next generation of carbon nanotubes engineered running shoes.

HI-TEC SHOES

Hi-Tec waterproof walking boots.

Hi-Tec synthesized the first walking boots to feature P2i waterproofing nanotechnology.

Innovation

- An innovative technology developed by P2i applies a waterproof polymeric material at the nanoscale which repels water under wet conditions and at the same time causes perspiration to evaporate.

Nanotechnology

- A patented plasma based technology that can coat objects with a permanent nano polymeric layer on every single fiber of the fabric.

Results

- Within only few months of the initial launch, the new hiking boots achieved a global sales record of over 200,000 pairs [3].

Drivers

Technology push. Although the technology was originally developed for military applications, Hi-Tec spotted it as an excellent opportunity to revolutionize the footwear market.

Market pull. The hiking and running shoes markets are continuously striving for new products that absorb less moisture, remain dry and are lighter in weight.

Competitive market. A highly competitive market in sports' footwear between tier companies drives the need for differentiation through innovative ideas.

Barriers

First entrant risk. Hi-Tec was the first company to introduce the P2i water repellent technology in its products, therefore, it had to face the risk of unknown market response.

Bridging the gap. P2i ion-mask technology developers are scientists and physicists with no expertise in footwear manufacture and business, thus Hi-Tec had to work hard to bridge this gap towards success.

Threat of new entrants. Although the technology required hard work to integrate it in the shoe business, as soon as other companies adopted it, Hi-Tec could lose a big share and the market would be highly competitive.

Enablers

R&D. Hi-Tec continuously invests in R&D and product innovation to lead the market with first entrant innovative solutions.

Environmentally friendly. The P2i process uses minimal amounts of chemicals with minimal waste, compared to waterproofing membranes, and operates at room temperature.

Build on existing franchise. The company managed to mitigate some of the first entrant risks through integrating the technology into its already highest selling boots.

Impact

Consumer satisfaction. The boots are 100 % waterproof and at the same time 100 % breathable because, unlike membranes, the molecules coat the fibers separately leaving the spaces in-between open.

Operational. The P2i waterproofing nanotechnology in the boots is less expensive than otherwise integrating waterproof membranes in a pair of shoes and was launched at the same cost of waterproof membrane shoes.

Market expansion. The success of the applied technology in Hi-Tec boots helped open new successful businesses through applying the same technology on gloves and hats.

What's Next?

P2i is attempting to apply the water repellent technology in the electronics sector thus serving a new and huge market.

EDDIE BAUER

Nanotechnology enters the Fashion world

Eddie Bauer licensed Nano-Care technology from Nano-Tex to develop stain resistant textiles.

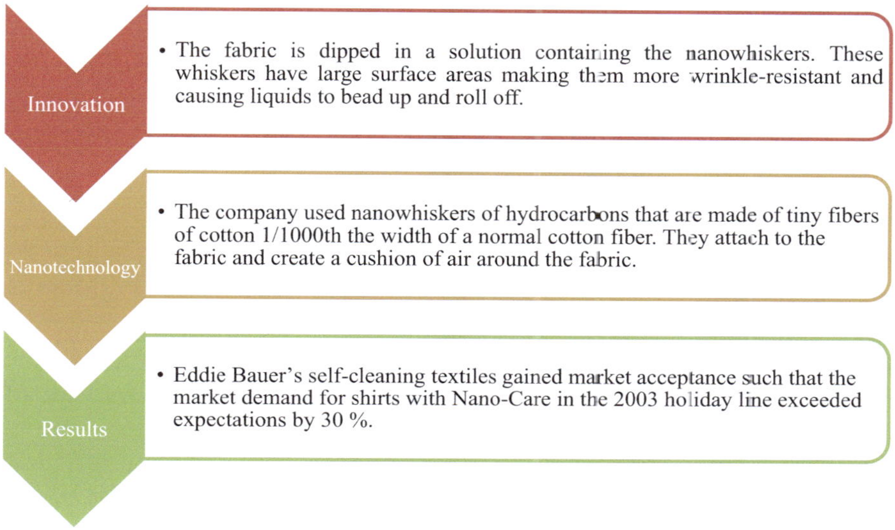

Drivers

Fashion market congestion. In a crowded fashion market, nanotechnology can be a lucrative point of differentiation driving fashionable brands to take first steps to incorporate nano-enabled features in their brands.

Market pull. Style conscious consumers with busy lifestyles drive the growing demand for new features in their clothing.

Market size. 16 % of the 55 billion dollars in US menswear sales in 2006 was on garments with some type of technological product enhancement whether wrinkle free or stain resistant [4]. This reveals the growing potential of such technologies in the market.

Barriers

Public fear. Limited public understanding and/or acceptance of textiles containing nanoparticles pose some risks because of the fear of nanoparticles' escape during cleaning or repeated wear or even during disposal.

Unproven durability. Lack of standardized evidence regarding the durability of nano-enabled textiles with repeated wash and wear posed some durability concerns especially that some shops advise consumers that the treatment in self-cleaning textiles will survive around ten washes only.

Slow adoption. Many consumers shop for image and style first and only slowly understand the added benefits of nano-enhanced textiles which causes a slow rate of adoption.

Enablers

First entrant. Eddie Bauer was the first retailer to pioneer the use of the self-cleaning technology in 2000 with the use of Nano-Care treatments in khaki pants.

Versatility. Nano-Care is compatible with cotton, wool, polyester, silk and rayon. The treatment is built in the fabric rather than coated onto it, thus it is permanently bound to the fabric with no risk of escaping.

Cost efficiency. Although treated pants and shirts have slightly higher prices than untreated ones, they could still be perceived as being cost effective by reducing washing bills.

Impact

Operational. The successful nanotechnology contribution to Eddie Bauer textiles lead the company to employ other nano-based features in their textiles like quick drying and moisture control.

Survive the battle. Nanotechnology supported Eddie Bauer's survival in the markets because it suffered fierce competitions from international producers—China—who had access to cheaper sources of material and labor.

New business concept. Through its nano-based textiles, Eddie Bauer enjoyed a new concept and a competitive advantage in its markets that was not based on price, but on differentiation.

What's Next?

The success of nanotechnology enhancements in fashion paves the road to compete in the niche markets of technical textiles.

ALEXIUM

Defenders need defense.

The US Department of Defense entered a strategic and commercial relationship with Alexium to develop nano-enhanced textiles for military applications.

Innovation

- Nanocoatings are applied around fibers via a "shrinkwrap" effect. Upon exposure to flames, the FR nanocoating on Nylon forms a nanocomposite char layer that retards flame and protects against dripping and melting.

Nanotechnology

- Reactive Surface Treatment (RST) is a patented cold microwave curing process that controls the nanoscale deposition of chemical functionalities on nylon fibers.

Results

- The success of the innovative process in military applications opened the door for the wide commercialization of the technology. Alexium and the US Air Force Research Laboratory in Florida signed collaborative agreements to develop and commercialize the patented technology.

Drivers

Market size. Nanotechnology continues to contribute to protection and military related textiles which own the lion share of the technical textiles market, predicted to further grow.

Regulatory push. The market demand for Flame Retardant (FR) fabrics is high, however, the most effective FR are halogenated chemicals which are strictly banned by the US and Europe's regulations due to their bioaccumulation and toxicity.

Customer demand. Although the fire resistant uniforms previously used by the US military were of high performance, they had high costs and low durability.

Barriers

Versatile but flammable. Nylon is a highly versatile and an inexpensive material that has many uses and applications especially in technical textiles but is inherently flammable.

Upscaling. The transition of the coating technology from laboratory scale to coating millions of meters of annually produced nylon is a challenge.

Disruptive technology. The technology has the potential to alter the current market with risks of slow adoption especially that the first customer is the US military who pursues extensive product testing and validation.

Enablers

Eco-friendly. The FR is non-halogenated thus meeting the stringent US and Europe's requirements. In addition the RST process is a low energy process with minimal footprint.

Huge investments. The US government invested over 30 million dollars to develop the technology and apply it in military applications.

Customer and partner. The US DOD is the first customer for the FR technology, which secured a stream of initial revenues for Alexium.

Successful validation. The technology passed several testing and validations conducted by Alexium and the US DOD.

Impact

Award. In 2009, the RST technology was awarded the platinum prize WBT *World's Best Technology* in the Annual Global Investment and Licensing Forum for Emerging Technologies.

Intrinsic feature. The enhancements are nanoscopic in range, thus the textile is light in weight with no significant change in the feel or the breathability of the fabric.

High profits. The technology was primarily targeted at the US military, focusing on around 1.4 million soldiers serving on active duty and 800,000 soldiers on reserve duty. The technology would annually modify at least four million linear yards per year achieving high profit margins and the DOD would gain nominal royalties on sales.

What's Next?

Alexium is working on expanding the FR technology application to non-textile fabrics such as carpets, furnishings and furniture and thus expanding into new markets.

RADIATION SHIELD TECHNOLOGIES

Light radiation protection gowns

Radiation Shield Technologies was able to fabricate special composite fabrics that contain nanoparticles to protect against radiation.

Innovation

- A radiation blocking light and flexible fabric that effectively protects against radiation exposure (same as lead shields).

Nanotechnology

- Nanostructured composites of radiation protecting nanoparticles of barium, bismuth or tungsten compounds dispersed in a plastic polymeric matrix to form films that are layered with fabric material.

Results

- Made effective radiation protection vests that are capable of reducing alpha and beta transmission and, to a lesser extent, gamma penetration.

Drivers

Military. Military personnel get in contact with bombs and nuclear weapons. As a result, they would need this suit. They would also require it to be light to be able to move freely in the battlefield.

Medical. Medical personnel working with x-rays and other radiations for imaging purposes of patient conditions must be protected by wearing suitable suits. The lead suits used to be very heavy and the presence of light weight flexible suits was always a necessity for ease of movement all day.

Barriers

High cost. The cost and price of such suits is extremely high due to the technology and material used in them. Price ranges from 2000 to 10,000 dollars per suit.

Enablers

Experience. Radiation Shield Technologies has been working for a decade now on innovating and designing radiation protection suits and fabrics.

Affiliations. The company is affiliated with a number of reputed companies in the field of radiation decontamination including Powerplus Cleaning Solutions for radiation disasters, CBRNE Systems for detection of explosives, Dose Gard for personal radiation dosimeters and the New York Iron Works for police wear.

Customization. According to the application, the Demron suits can be customized to resist chemicals, corrosives and biological threats as well as radiation.

Regulations. Government regulations that require people working in radiation exposure areas to wear such protection suits are also some of the major enablers.

Impact

Due to the company's strong market presence, it was capable of patenting its technology in several countries.

What's Next?

The company can make radiation covers for equipment whose performance can be affected by nearby radiations. It can also develop the suits to be sort of smart suits and enable digital components to be integrated in them.

DUPONT STRETCHABLE CONDUCTOR

Smart computerized shirt

Dupont was able to fabricate stretchable conductive wires that are invisibly integrated into clothing.

Innovation

- DuPont developed stretchable electronic ink materials for smart clothing and other wearable electronics.

Nanotechnology

- It used a conductive paste made from silver nanoparticles that have excellent stretchability, adhesion and conduction. The paste is encapsulated to withstand washing.

Results

- DuPont was able to develop wearable electronics that are designed to deliver stable performances despite repeated elongation. Smart clothing enabled by DuPont inks is washable and durable—it can withstand up to 100 wash cycles. The stretchable circuits are seamlessly fused with standard fabrics allowing for unprecedented comfort and freedom in wearable electronics design. The stretchable ink is compatible with a wide variety of fabric and film substrates as well as fabrication processes.

Drivers

Data acquisition for athletes during training. Monitoring devices for the vital signs of athletes doing exercises need to be lighter in weight and to not be bothersome to athletes during training.

Tracking the position of soldiers during battles. The ability to track soldiers in the battlefield without burdening them with heavy tracking devices is necessary as they already have to carry heavy weapons and be swift during battles.

Health monitoring in patients. Patients need lighter monitors to be attached to them so they can freely turn over in bed, especially bedridden patients. This can be solved by the use of wearable electronics.

Barriers

Safety. The integration of circuits and wiring without any electric shocks or adverse health effects is challenging. However, the encapsulation technology made by DuPont has enabled the isolation of the flexible electronics from the bodies of the wearer.

Ease of fabrication. It is not easy to fabricate such a technology, however, DuPont's two century experience and strategic direction of spending on research to provide advanced materials allowed it to fabricate such smart clothing.

Durability. DuPont's clothing achieved the ability to withstand washing and stretching through the encapsulation technology and stretchable conductive pastes.

Enablers

Advancement of the Technology. The advancement in nanotechnology and in polymer science has enabled the design of flexible conductives made from dielectric polymers, silver and other metal nanoparticles.

Experience and long standing market presence. DuPont has been in the market for over two centuries designing high quality fibers and fabrics and gaining experience. It has also allocated intensive research in new technological areas like nanotechnology. This earns the company its place as the world's most innovative science company because it was capable of turning scientific breakthroughs into commercial breakouts.

Collaborations. DuPont made many collaborations on the wearable electronics products that it made including collaborations with North Carolina State University, and the National Science Foundation Nanosystems Engineering Research Center on Advanced Self Powered Systems of Integrated Sensors and Technologies.

Impact

A complete product profile was developed that enables attachment of sensors, touch screens, cameras and other electronic options. Products are not only used for wearable clothing but also for textiles, décor or flexible circuit boards.

What's Next?

The next step for this technology would be designing electronic wear to be used in hospitals that, not only detect health vital signs, but also administer doses of medication in response to a patient's status.

References

1. Report on nanotechnology and textiles (2010) ObservatoryNANO
2. Fenn D (2007) Sports market review. http://designbusiness.wikispaces.com/file/view/Sports+Market+Keynote.pdf. Accessed 21 Aug 2015
3. (n.d.) http://www.hi-tec.com/id/news/51/sales-kendal/. Accessed 21 Aug 2015
4. Vella M (2007) High style goes high tech. BloombergBusiness. http://www.bloomberg.com/bw/stories/2007-02-05/high-style-goes-high-techbusinessweek-business-news-stock-market-and-financial-advice. Accessed 24 Aug 2015

Chapter 10
Sports Equipment

Nanotechnology greatly contributed to the growth of the sports market by making sports equipment more popular. This chapter shows how nanotechnology could add enhanced features to sports equipment by making it lighter and stronger. The chapter will further discuss how the sports equipment market is enjoying a unique privilege—that the consumers are athletes whether amateurs or professionals—over other markets. This privilege encouraged sports equipment manufacturers to invest in nano based equipment. Consumers are usually well educated and health-conscious, so they are willing to pay more for better performing equipment. Nanomaterials are to known to add unique features to sports equipment and many of the nano-enhanced equipment has granted better performances to athletes in worldwide competitions and in Olympic games. The latter case, however, is still under debate because nano-enhanced equipment is more expensive and not all athletes can afford it, meaning that not all athletes have equal opportunities in competitions.

Introduction

Sports equipment researchers are continuously developing and improving sports equipment. Nanotechnology is the most convenient and effective technology that can be employed to enhance sports equipment performances. The sports equipment market serves a certain segment of well-educated consumers who are especially conscious about their health. Therefore, such consumers are highly willing to pay extra costs for better performing sports equipment.

Sports equipment manufacturers who decided to develop their products using nanomaterials used a smart marketing strategy by introducing their nano enhanced products into Olympics and having them used by famous athletes. This facilitated the rapid adoption of nano-enhanced products by consumers. The market potential of

M. Anis et al., *Nanovate*, DOI 10.1007/978-3-319-44863-3_10

sports equipment is expected to reach 72.8 billion dollars by 2016 [1]. Thus, there is huge potential for nanomaterials to be incorporated in sports equipment.

Scientists have found many applications of nanotechnology in sports equipment. They managed to produce stronger and lighter weight equipment. Minute changes in sports equipment using nanomaterials produce significant positive changes in performances and reduce the incidence of injuries. These minimal changes are major contributing factors in winning or losing a competition.

Cases

EASTON BIKES

Superlight and strong bicycles

Easton bicycle division partnered with the nanotech company Zyvex to strengthen the resin used in its bicycle components.

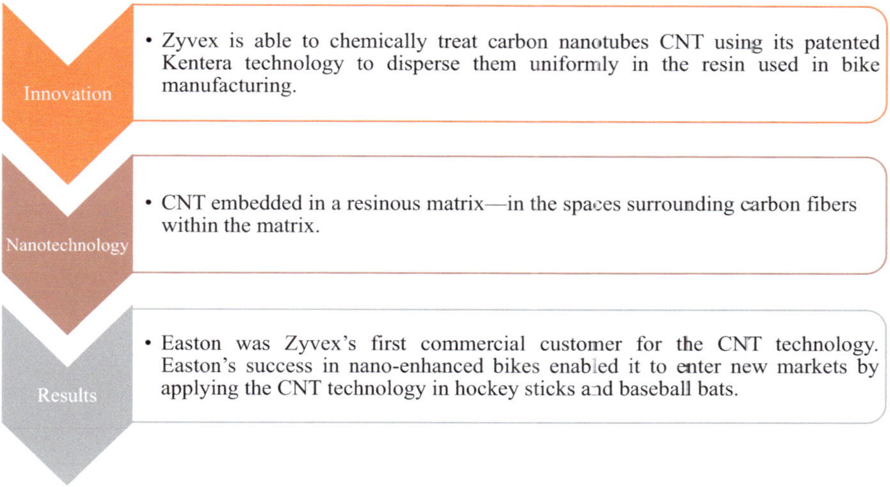

Drivers

Technology push. Nanomaterials can add unique enhancements in the weight or strength of bikes as the slightest changes can boost performance.

World level championships. The Olympics and the World Cup have made the sports equipment industry highly popular and have stimulated the incorporation of new technologies.

Market size. CNT market is expanding with sporting equipment constituting around 14% [1] of its annual consumption. This figure is expected to grow even more in the near future.

Barriers

Disruptive technology. Ever since the discovery of carbon nanotubes' in 1990, many advancements have been achieved to optimize their synthesis and price, yet they are still too expensive and complex to deal with.

Higher end price. The nano-enhanced bikes are commercialized at upper-end prices which makes them unaffordable by many athletes in competitions. Many call for restricting nano-enhanced equipment in competitions to be able to judge the results fairly.

Technical hurdles. As an early adopter of Zyvex's technology, the company faced many challenges in its attempt to optimize the best performing dispersion conditions for the CNT in the epoxy resin of bicycles' components.

Enablers

Skilled partner. Zyvex, Easton's CNT technology supplier, is claimed to be the first molecular nanotechnology company that worked on developing the technology with NASA.

Entrepreneurial initiative. Easton was one of the earliest companies that decided to with experiment nanotechnology to improve its equipment.

Nano privilege. CNT, as a nanostructure, are as stiff as diamonds and 100 times stronger and six times lighter than steel which makes them the ideal nanostructure for improving bikes.

Impact

Ahead of the game. With the launch of the world's first nano-enhanced bicycle, Easton held a leading position in the market and achieved good sales and trust from champion road racing cyclists.

Competitive advantage. The CNT imparted 10–20% [2] higher toughness and strength to the bike frames. At the same time, they had much lower weights and longer life spans compared to traditional bike materials.

Swiss customer. BMC, a leading Swiss Bicycle manufacturing company, is the first company to employ Easton's CNT technology in the frames of its Pro-machine bicycles and won the gold award for design in the Eurobike show.

What's Next?

More stiffness at lighter weights is the magical solution to many applications in the automotive and aerospace industries, thus providing potential markets for the technology.

MARUMAN GOLF CLUBS

Golf clubs with stronger impacts

The Japanese golf equipment company Maruman & Co. uses nanotechnology to enhance its product lines.

Innovation	• The nanoparticle developed by Maruman is made of titanium which adds great strength to the drivers and at the same time great flexibility.
Nanotechnology	• The company used Fullerene titanium, a high stiffness nanomaterial.
Results	• Maruman's nano enhanced *Majesty* series is sold at premium prices to exclusive golfers. The series is prized and was listed in the Forbes magazine as the most expensive set of golf clubs worldwide.

Drivers

Market potential. The global market for sports equipment in 2011 was valued around 64.9 billion dollars and is expected to reach around 72.8 billion dollars by 2016 [1].

Market trend. Nanotechnology is penetrating different markets with real life added benefits in many industries. The sports and leisure market is a valuable market opportunity for nano-enhanced products.

Innovation strategy. For over 35 years, Maruman has been engineering its golf equipment with the most updated technology innovations and is the first to apply nanotechnology in golf equipment.

Barriers

Governing bodies. There are no rules that govern nanotechnology in sports equipment. Many of those opposing the technology are afraid that nano-enhanced equipment would mask the actual athletic performance which may lead to a complete ban to nano-enhanced equipment in official competitions.

High expenses. Because nanotechnology is very new in the field, its implementation in golf clubs required Maruman to build its own know-how which was costly. Additionally, the processes to manipulate metals at the nanoscale were expensive.

Most golfers are traditionalists. Nanotechnology is a developing field and many golfers are not aware of its significant contribution in products. This poses risks such as consumer fear of anything nano and slows the adoption of nano-enhanced products.

Enablers

Premium consumer. Golf is a luxurious sport. Golfers are accustomed to paying premium prices for better and more prestigious golf equipment.

Design and analyses. Maruman has a long history of over 20 years in using computer aided design systems and Finite Element Method (FEM) stress analysis to precisely design the thickness of its drivers' heads.

Intellectual golfing. Maruman's commitment to innovation and to the implementation of the latest technologies led it to develop a unique titanium called Fullerene Titanium. The development is patented and added to the company's full library of patents.

Impact

Intrinsic features. The driver face is ultrathin and hard to deform producing stronger, lower weight golf clubs that can propel balls farther (15 extra yards).

International presence. Maruman's innovative golf equipment enabled it to successfully expand its presence in Europe and North America through participating in official golf shows and the US Professional Golf Tours.

Internal. Since the launch of its nano enhanced products, Maruman achieved great sales records on local and international scales.

What's Next?

Maruman & Co. is continuously observing the recent technological trends to improve its product lines and introduce innovative versions from its existing golf clubs series.

WILSON NCODE TENNIS RACQUETS

Better serves with Wilson nCode series

Wilson, a leading company in sporting goods, introduces nanotechnology in its tennis racquets.

Innovation

- Tennis Racquets are normally made of billions of woven carbon fibers that contain microscopic spaces in between. These gaps act as weak points that compromise strength. Thus, Wilson attempted to fill these gaps with nanoparticles to add extra stiffness and stability.

Nanotechnology

- The company used Silica nanoparticles.

Results

- In less than 3 months after Wilson introduced the nCode series, the racquets became the bestselling in the high end markets.

Drivers

Market demand. To fulfill the market needs, tennis racquet manufacturers are always aiming to manufacture tennis racquets that will retain their stiffness and will not become soft with time.

First choice by top players. Wilson's racquets has maintained a strong position among the world's top tennis players since 1949. This coerces the company to invest in developing the quality and performance of its equipment to keep its position.

Sporting chance for nano. According to the Tennis Industry Association, the sales of tennis racquets have been down for over a decade, which presents an opportunity for early adopters of nanotechnology in their equipment to re-attract consumers and revive the market.

Barriers

High cost. Nanotechnology in sports equipment is expensive and challenging. Retailers are slowly introducing nanotechnology enhanced products to observe first their market acceptance and adoption.

Consumer learning. Most retailers are unaware of the concepts and applications of nanotechnology in sports equipment. This requires manufacturers like Wilson to work on enhancing the learning of its customers.

Consumer decision. Nanotechnology enhancements are invisible and consumers' buying decisions are affected greatly by visible features.

Enablers

Long history. Wilson was one of the early introducers of nanotechnology in its golf equipment as it launched it in 2004 and ever since, it has been developing and expanding its application in all its product lines.

Market leader. Wilson is a global market leader with 25 % share in global tennis balls and 36 % share in global tennis racquets. It invests heavily in the latest technologies to maintain its position.

Endorsements. Wilson attempts to increase the popularity of its newly introduced racquets by outlining and marketing that each series is the premium choice of one of the top tennis players.

Impact

High performance. The nCode series proved to be 22% [3] more powerful, twice as stable and as strong as comparable tennis racquets in the market.

Premium user. Wilson's nCode series achieved, through nanotechnology, great enhancements that they became the racquets of choice by number one men's player Roger Federer at Wimbledon 2009.

Product line development. The success of the nCode series in the market led Wilson to further enhance and optimize them and introduced the K series in 2007 followed by the BLX line. All are employing nano-enhanced features.

What's Next?

Wilson is investigating the use of new materials that are cheaper which will enable the company to reduce the costs of the nano-enhanced racquets and increase their revenues.

References

1. (n.d.) http://www.nanowerk.com/spotlight/spotid=30661.php. Accessed 15 Sept 2014
2. (n.d.) http://www.bike-eu.com/home/nieuws/2006/3/nanotechnology-moves-into-bicycles-1017788. Accessed 17 Sept 2014
3. (n.d.) http://www.tennisexpress.com/info/wilson-ncode-technology.cfm. Accessed 20 Sept 2014

Chapter 11
Nutraceutical Industry

With the increased costs associated with diseases and their treatment, nutraceuticals are being increasingly adopted by different consumers. Nutraceuticals represent a certain class of dietary supplements that are able to achieve both health and medicinal benefits. However, some nutraceutical ingredients suffer from poor delivery, insolubility and undesired side effects. The introduction of nanotechnology within the nutraceutical industry allows it to overcome such problems. Thus, several nano-based nutraceutical products already exist in the market. Nevertheless, safety concerns have risen concerning such nano-based nutraceuticals because, as the material decreases to the nanoscale, its ability to diffuse into different parts of the body, including the blood brain barrier, increases. Thus, this necessitates the conducting of intensive safety studies for these products. It should be noted, however, that the lack of regulations concerning nano-based nutraceuticals has allowed easy entrance of different products into the market.

Introduction

Nutraceuticals is a class of dietary supplements that is able to achieve both medicinal and health benefits. Nutraceuticals are able to provide the main nutritional values of food in addition to treatment and prevention of diseases to some extent. Nutraceuticals are mainly used in functional foods and dietary supplements. Since the 1990s, there has been a considerable consumer interest, especially in developed countries, towards healthy diet and nutraceuticals in an attempt to reduce the expensive costs associated with diseases and their treatment [1].

The global nutraceutical industry reached 50.4 billion dollars in 2010 [1] and is expected to reach 204.8 billion dollars by 2017 with a compound annual growth rate of 6.3 % [2]. However, nutraceutical ingredients suffer from poor delivery characteristics where

M. Anis et al., *Nanovate*, DOI 10.1007/978-3-319-44863-3_11

only a fraction of the administered dose actually gets into the system and exerts its required effect. In a multivitamin pill, for example, it was found that only 45 % of the people are able to break the administered pill fast enough to absorb its ingredients into their system [2].

Nanotechnology is now enabling the delivery of hard to deliver nutraceutical ingredients like botanical extracts, fat soluble vitamins, nutrients and peptides into their required sites and in the required doses [3]. Nanoemulsions and polymeric micelles based delivery systems were found to enhance the bioavailability of nutraceutical ingredients. In nanoemulsions and nanomicelles, the compound of interest is entrapped within a lipid based delivery system. Such approach was found to enhance the thermodynamic and kinetic stability and the ability to deliver the required nutraceuticals through cell membranes as well as the feasibility of combining both hydrophilic and hydrophobic ingredients within the same formula [4]. Moreover, it was found that such delivery systems enhanced the oral bioavailability and the efficacies of the delivered nutraceuticals [4].

Notably, though, there are safety concerns associated with nano-based nutraceuticals. As an ingredient decreases in size to the nanoscale, there is an increase in its surface to volume ratio, which in turn increases the material's solubility, bio-kinetics and absorption. Thus, upon formulating a nano-based nutraceutical, the ingredient optimum nano-size concentration absorbed as well as other related safety studies must be well established [2].

Cases

FRUTAROM FLAVORS FOR FOOD AND BEVERAGE

Oily flavors in aqueous media without two phase separation; nanocapsules of flavors for the food and beverage industry.

Through the utilization of nanocapsules, oily flavors and ingredients were able to be incorporated in aqueous media.

Innovation

- Clear and stable integration of (lipophilic) oily ingredients and flavors into aqueous beverages without the occurrence of two-phase separation.

Nanotechnology

- The nanotechnology used were nanocapsules. This technology allowed functional lipophilic (oily) ingredients like catechins, CoQ10, lycopene and phytosterols to be integrated into an aqueous phase.

Results

- Nanocapsule based flavors are currently marketed worldwide by Frutarom in the United States.

Drivers

Consumer Demand. Since the past decade, consumers of developed countries have established an increased tendency and focus towards health and wellness [5, 6].

Governmental Support. USDA and FSA support research and development of functional food, nutrient delivery systems as well as methods of optimizing food appearance [6].

Strong Market. Nanotechnology is progressing in the food and drink industry which necessitates that all companies, should they wish to thrive in such a market, stay on top of this dynamic development [1].

Barriers

Scaling up. Scaling up the manufacturing process is a difficulty facing nano-enabled technologies as it requires special characterization techniques [7].

Nano-based R&D. The required expertise and infrastructure for nano-based research and the required development is extremely challenging [7].

Public Acceptance. Public acceptance still represents a major hurdle against nanotechnology inclusion within the food industry [7, 8].

Enablers

Increased Nano-based R&D. In the previous decades nano-based research has increased, especially in developed countries [7].

Partnerships. A joint venture between Hebrew University of Jerusalem (technology developer; Nutralease) and Frutarom (technology marketer) has enabled the commercial success of such an innovation [7, 9].

Buyer pull strategy. There has been recent increased demand for healthier foods. This demand increase has been a major factor in the increased consumer acceptance for nano-based products.

Impact

Internal. The developed technology extended into a number of products. It has also enabled the chances of strong partnerships with major industries like the Shemen industry.

Industry. The technology developed has also been extended to cosmetics and the delivery of essential oils and drugs. It has been distributed in the United States, Mexico and Canada [8].

Consumer. Nanotechnology enhanced the developed products by allowing the incorporation of oily phase in aqueous phase in a clear solution with no phase separation. Moreover, through Nutralease, innovation junk food can be turned into health food [7].

What's Next?

The developed nanocapsules could be further used in other food and dairy industries for inclusion of oily ingredients in aqueous media and vice versa. Furthermore, such nanocapsules could be used in cosmetics for inclusion of vitamins and other ingredients in the cosmetics thus increasing their bioavailability and improving the consistency.

SHEMEN INDUSTRIES: CANOLA OIL

Canola oil that is actually able to decrease cholesterol level

Nano-Sized Self-Assembled Structured Liquids (NSSL)—minute-compressed micelles—used in the canola active oil industry. Such technology allowed the incorporation of insoluble phytosterols in the oil which allow reduction in the intake of cholesterol.

Innovation

- Shemen Industries, in its canola active oil, was able to integrate insoluble phytosterols in the produced oil. Phytosterols inhibit the transportation of cholesterol from the digestive system to the bloodstream, thus lowering the cholesterol level and decreasing the risk of heart disease and stroke.

Nanotechnology

- Nano-Sized Self-Assembled Structured Liquids (NSSL) which are minute-compressed micelles are used in Canola Active Oil. The nanoscaled micelles contain and solubilize the phytosterols to increase their bioavailability.

Results

- The developed technology allowed the oil to reduce cholesterol intake by 14%.

Drivers

Consumer Demand. Over the past decade, the major food and beverage companies have undergone fundamental strategic shifts to take into account the increasing consumer focus on health and wellness [5, 6].

Investment reduction versus strong market. The technology was already developed by the Nutralease startup company and success was achieved in the products adopting the technology. Thus, no new investment was required by Shemen Industries to be spent on R&D.

Decrease in company's profit. In 2001, the worldwide crisis in high-tech and capital markets has earned Shemen Industries control of about 47 % of the sales of the edible oil market to achieve low profits of less than 3 % [10].

Barriers

Science philosophy debate. Debate around nanotechnology is similar to the debate around Genetically Modified Organisms (GMOs) in being a new technology with benefits to the product but at the same time not accepted by the consumer due to fear of side effects.

Lab to market transition. Transition from the lab to the marketplace is a long and tortuous path that requires intensive funding, collaboration, patenting, licensing, regulatory approval and consumer acceptance [7].

Food technology neophobia. "Better safe than sorry" is the consumer's perception about nanotechnology [11]. Food technology neophobia affects perceptions of the benefits of nanotechnology.

Enablers

Developed technology and partnership. The technology used by Shemen Industries in its canola oil was already developed and applied by Nutralease. This was a huge enabler for Shemen, achieved through partnership with Nutralease industries.

Lack of regulations. In January 2012, the FDA was sued for lack of regulation of existing products on the market.

Buyer pull strategy. The recent increase in demand for healthier food has resulted in an increase in consumer acceptance of nano-based products. The developed nanocapsules have allowed the consumption of oil while decreasing levels of cholesterol and heart disease [7].

Impact

Internal. The nano-based products allowed the company to face the decrease in profits that occurred in 2001. In 2001, the company decided to adopt a new marketing strategy which depended on making consumers perceive the new products offered by the company from a quality point of view rather than from a price point of view. This enabled it to acquire new market shares for its products [10].

Industry. The canola nano-based oil developed by Shemen was regarded as one of the top five nanotechnology products by Forbes in 2005.

Consumer. The consumer can enjoy using oil without having to fear the risk of cholesterol or heart disease [12].

What's Next?

The present innovation allowed the inclusion of beneficial nutrients, ingredients and supplements that face problems with solubility. It can be used to convert normal food into nutraceuticals; i.e. deliver vitamins and medications through the food we eat. For example, multivitamin ice-cream can be developed for children. Moreover, such innovation can be extended to the cosmetic industry allowing two immiscible components to be incorporated together homogeneously.

AQUANOVA WEIGHT MANAGEMENT FORMULA

Two products for weight management in a single dose

Aquanova, a supplier of liquid formulas, has developed nano-micelles for the simultaneous delivery of active ingredients for weight management in a single dosage form.

Innovation

- Aquanova has developed a new technology—NovaSol technology—that combined two active substances for fat reduction and satiety into a single nano-carrier and thus a single dosage form. The new approach is used for intelligent weight management.

Nanotechnology

- Nanomicelles of an average of 30 nm in diameter allowed the delivery of CoQ1O to address fat reduction and alpha-lipoic acid for satiety together in a single carrier.

Results

- Due to the success of the NovaSol technology, it was also tried and developed to create other products like vitamin E preparation that does not cloud liquids and a vitamin C preparation. Moreover, Aquanova recently won the Beverage Award 2009 as "best new ingredient" for its new NovaSol enabled product line. In 2006, Aquanova received the "Excellence in Technology Award".

Drivers

Consumer Demand. In developed countries the food industry is currently driven by the need for fresher and healthier food [6].

Improving ingredients efficacy. Usually the body converts the nutrients into micelles before it can use them. Thus, if the ingredients are delivered already miscelled, as was done through Aquanova's nano-based weight management formula, this cuts down the amount of ingredients that would otherwise be lost in the conversion process [13].

Product differentiation. The company's strategy to develop the product was to achieve product differentiation and avoid the hype surrounding other weight loss products in the market [13]. The nano-based technology enabled the company to develop, for the first time, a natural combination for weight management and weight maintenance in a 30 nm micelle allowing for better delivery to the body.

Barriers

Safety. If nano-nutritional additives and supplements provide an excessive dose of some vitamins or nutrients, these may have a toxic effect or interfere with the absorption of other nutrients. Nano-agglomerates is another problem that may arise from nano-based supplements [14].

High cost R&D. The development of Aquanova's nano-based weight management formula was funded by Aquanova itself in a six figure R&D budget [13].

Public acceptance. A study on public acceptance of nanotechnology in foods and food packaging found that it was more accepted to include nanotechnology in the packaging than to include it in the food. However, in both cases participants were hesitant to purchase the product [8].

Enablers

Strong scientific foundation. The strong scientific foundation of Aquanova allowed it to develop the NovaSol technology in a six digit R&D process according to corporate development manager Frank Behnam [13].

Partnership. Aquanova's partnering with Degussa to develop a soluble and bioavailable form for one of the ingredients resulted in the launching of a new product. Moreover, Aquanova used the technology to create a vitamin E ingredient that does not cloud liquids and a vitamin C in partnership with BASF [13].

Buyer-pull strategy. The company expected that the product could be particularly well received in the US where weight management is a big issue.

Impact

Internal. The NovaSol technology used by Aquanova was extended to other products produced by the company like SoluE and SoluC [13]. Just recently Aquanova used this technology to formulate a soluble form of the magical ingredient curcumin [15].

Market shares. The NovaSol technology allowed Aquanova to acquire new market shares in strong markets like functional food and nutraceuticals markets.

What's Next?

The developed NovaSol technology could be easily integrated into different food products converting them into functional foods and nutraceuticals, i.e. deliver vitamins and medications through the food we eat. The NovaSol technology could also be used in cosmetics and pharmaceutical industry as in drug delivery and impregnated wound dressings.

RBC LIFE SCIENCES: CHOCOLATE SHAKES

Chocolate that does not increase weight!

RBC Life Sciences, through the use of nanoclusters, was able to introduce chocolate shakes suitable for weight management.

Innovation

- RBC Life Sciences introduced its chocolate product that allowed enhanced cocoa flavor without the need for excess sugar. This strengthened the sweet taste of chocolate without the associated weight gain.

Nanotechnology

- The used technology constituted nanoclusters of nanosized silica particles covered with molecules of chocolate flavor. The increase in the surface at the nanoscale allowed for increased flavor without the need for excess sugar.

Results

- Through the utilization of the new nano-based technology, RBC life sciences was able to develop a new line of skincare and nutritional supplements and managed to internationally expand and create new distribution channels.

Drivers

Demand Pull. There has always been a strong need among consumers to enjoy sweet food without the fear of excessive weight gain and other health risks due to the presence of extra sugar [7].

Growing market. Nanotechnology will continue to progress within the food and drink industry, especially in the developed world. In order to stay within this dynamic competition and progress, companies need to be part of the nanotechnology revolution [1].

Increasing cost of health care. The high cost of healthcare especially in developed countries was an important driver for consumers towards dietary supplements and healthy food to help them avoid health issues and related cost [1].

Barriers

Safety. Concern about the possibility of migration of the silica nanoparticles to inaccessible parts of the body has always been a major concern [16].

Required expertise, technology and cost. To reach such innovation, costs are required for obtaining the required expertise and technology [1].

Public acceptance. Public acceptance of nano-based food still represents a major hurdle against the widespread use of nanotechnology in food products [17].

Enablers

Science push. The increased research as well as application of nanotechnology in the nutraceutical and functional food fields has enabled different companies to increasingly adopt and implement the technology.

Strong R&D. RBC Life Sciences possessed a strong R&D center that enabled it to introduce nanotechnology solutions to many of its products including the chocolate shake [18].

Maneuvering FDA approval. RBC benefited from the confusion over the existing regulations of nanomaterials as it depended on silica being Generally Regarded as Safe (GRAS) by the FDA and that no new regulation existed for a material when its size decreased to the nanoscale [19].

Impact

Internal. After the success of its chocolate shake product, the company launched a new line of skincare and nutritional supplement products.

Market operations. Attracted by the new line, new distributors joined RBC and created new markets [19].

International expansion. A 10 year agreement was signed by the company that allowed its licensee in the former Soviet Union to continue. Moreover, the company expanded in new countries like Eastern Europe, Finland, Israel and Greece [19].

What's Next?

The technology developed by RBC Life Sciences allowed the use of less loaded material—chocolate—to achieve the same desired result—flavor. This was made possible by the increased surface-to-volume ratio of silica nanoparticles. Such technology can be applied to different fields like the pharmaceutical industry and the food and beverage industry. It can also be applied to the fragrances added to detergents used for clothes, decreasing their added concentration and increasing their penetration.

O'LALA CHOCOLATE GUM

Finally chocolate chewing gum!

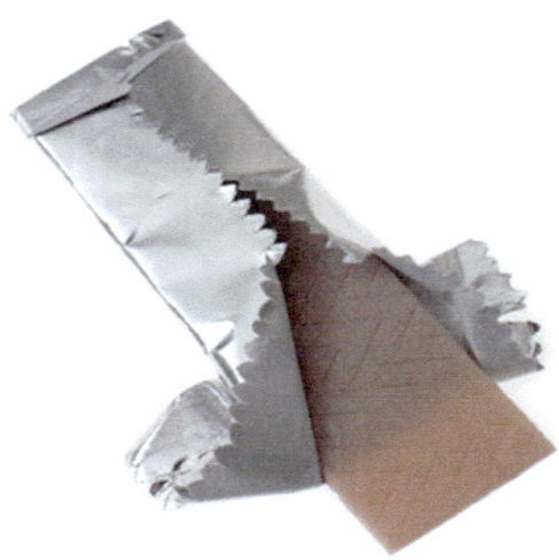

Nanotechnology enables the introduction of the first chocolate flavored chewing gum.

Innovation

- For the first time O'lala was able to produce chocolate chewing gum since usually fats found in cocoa butter cause the chewing gum to fall apart.

Nanotechnology

- The technology involved used nanoscale crystals of cocoa butter that allowed or modified surface morphology and resulted in a gum with a creamier texture and chocolaty flavor.

Results

- The first chocolate gum was finally produced. Such chocolate gum was developed to be sugar free and fat free thus making it a chocolate alternative for those trying to maintain their weights and for diabetic patients. The product was selected as one of the top ten products according to Forbes.

Drivers

Unmet challenge. Chocolate gum was usually impossible to produce since fat breaks the bonds required to hold the gum together.

Market opportunity. Implementing nanotechnology allowed utilizing decreased concentrations of cocoa clusters to present new market possibilities like markets of diabetic consumers and markets with consumers regulating weight.

Strong market. Adding health benefits to chocolate gum through its low sugar and fat content allowed it to enter into the market for functional foods, which has been witnessing a strong rise in the past few decades [20].

Barriers

Research and development. Extensive research is required to allow developing a product that delivers creamy low fat flavors and at the same time has the same mouth feel and taste as high fat products.

Venture financing. O'lala started as a small startup; its source of financing relied mainly on persuading venture capitalists to invest in the company [20].

Public acceptance. As with all nano-products of food, public acceptance is a barrier that will be overcome solely through consumer education.

Enablers

Nano-trend; nano-research. The increase in the nano-based research in the last decades made it possible to modify the high fat and sugar properties of chocolate.

Collaboration. O'lala collaborated with Ford Gum & Machine Co. to help manufacture O'lala's product concept.

Impact

Internal; line of novel products. After O'lala witnessed its success in gum, it utilized the same technology to produce other innovative food and beverage products to overcome problems that have always existed in these products like taste, efficacy and bioavailability [21].

Partnerships. After the company's gum production and success, it was able to establish a relationship with Sugar Mountain Candy, one of Canada's largest independent candy retailers [24].

Market. Since the production of O'lala gum, the product has had sustained and substantial growth. Its market presence was strengthened by establishing relations with a large number of retail outlets [21].

What's Next?

The technology developed by O'lala allowed the conversion of high fat and sugar content foods to healthier foods with low fat and sugar content. This sheds light on the fact that such a technology can be successfully integrated in the food oil and margarine industry as well as the desert and sweet industries.

ADVANCED SPORTS NUTRITION: CREATINE

Creatine with no side effects!

Advanced Sports Nutrition (ASN) introduced creatine without the associated side effects. This was achieved via the reduction of the size of the ingredients to the nanoscale.

Innovation

- ASN offered sports drinks with ingredients that are at the nanoscale. The nano-enabled product was able to overcome side effects associated with creatine like stomach upsets and headaches.

Nanotechnology

- The nanotechnology involved the inclusion of ingredients prepared as nanoparticles, thus maximizing uptake. It also decreased side effects due to reduction of the administered dose.

Results

- Market lead was achieved with its nano-based sports products. ASN's nano-enabled creatine product is the first product on the market with creatine as a main ingredient. The fact that it is at the nanoscale helps overcome the problems usually encountered with creatine.

Drivers

Demand pull. The problems that were associated with creatine administration were so troubling to the point that many athletes stop taking it despite its benefit [22].

Market lead. Developing a creatine that is devoid of the problems normally encountered with creatine will allow the company to be the first in the market to develop such a product and achieve market lead in such an important market [22].

Cost effectiveness. The company realized that nanotechnology and developing nanomaterials was more expensive. On the other hand, it noted that on the long term, it is more cost effective as it would be paying less and for a better result.

Barriers

Research and development. A high standard process of research and development is required to optimize a successful product, especially when this product is developed for human consumption.

Legislative hurdles. Due to absence of clear regulatory standards, the industry is not encouraged enough to develop nano-based food or food related products, especially with the presence of many consumer groups who are often criticizing nano-based products and raising the alarm of the public.

Public acceptance. Being acceptance by the public is still a main barrier against the development of new nano-products within the food sector. This is mainly due to the little information available about nanoparticle properties as well as their potential impact [17].

Enablers

Company and lab cooperation. ASN cooperated with different laboratories who were considered among the most reputable laboratories in the world to discover how to overcome the side effects associated with taking creatine [22].

Market need. The patients' needs to overcome the drawbacks they encounter with other creatine products in the market is a buyer pull strategy that enabled the marketing of the newly developed product and making it free of such drawbacks.

Strong scientific and industrial foundation. ASN is the only company in the world in that field that has exclusive access to the most advanced processing technology and purest ingredients [23].

Impact

Internal. The nanotechnology used by the company in its nano-enabled creatine product was extended to multiple products produced by the company.

Consumer. The product greatly improved the bioavailability of creatine and overcame the downsides encountered by consumers.

Market lead. The nano-enabled creatine product allowed ASN to achieve market lead with its development of the first creatine product in the market free of creatine's usual disadvantages [23].

What's Next?

ASN prepared the main ingredients of its sports drinks in the nanoscale which increased their bioavailability, thus decreasing the administered dose and decreasing the accompanied side effects. Such technology could be extended to drugs and cosmetics. It could also be extended to food components so as to enhance their effect and decrease their added concentration. Such food components include food coloring, artificial flavors, salt and sugar.

SHENZHEN TEA

Tea with enhanced flavor and properties.

Shenzhen reduced the size of the tea particles to the nano range resulting in tea with an enhanced flavor and an enhanced selenium supplement.

Innovation
- Nano-tea with enhanced flavor and enhanced supplement of selenium.

Nanotechnology
- Tea particles decreased to the nanosize. The nano-tea enhances the release of all the excellent essences of tea. Moreover, it boosts the adsorption. It also acts as an excellent supplement of selenium increasing its concentration by about ten times.

Results
- Shenzhen Tea was considered as the first nano-tea with enhanced properties.

Drivers

Science push. Nanotechnology research and nano-enabled products have witnessed a massive increase in the last few decades. This nano-trend represented a science push in different fields.

Improving ingredients efficacy. Decreasing the size of the tea to the nanoscale increases its bioavailability and absorption. It also enhances the flavor and provides a better supply of selenium.

Product differentiation. Developing a nano-tea will allow Shenzhen to achieve product differentiation among the numerous kinds of tea products in the market.

Barriers

Safety. The enhanced absorption associated with the small size can lead to an excessive dose of certain nutrients which can be toxic or interfere with the adsorption of other nutrients. Moreover, nano-agglomerates represent another problem that may face nano-based foods.

Infrastructure and R&D. Advanced research, expertise and infrastructure are necessary requirements for the optimization and implementation of the new technology.

Learning process. A good learning process is required to advertise the enhanced properties of the nano-based tea to achieve the competitive advantage of the newly developed product.

Enablers

Increased nano-research and products. The increase in research and nano-based products, especially in the functional food field, allowed Shenzhen to develop its nano-tea.

Lean regulatory guidelines. The confusion over the existing regulations concerning nanomaterials allowed the approval of nano-tea.

Impact

Internal. Acquiring the expertise and technology that allowed the development of nano-tea could open the door for Shenzhen to extend it to other products.

Market lead. Having introduced the first nano-tea in China, Shenzhen achieved market lead in this sector.

Consumer. The enhanced flavor and the increased supply of selenium resulted in consumer satisfaction.

What's Next?

Shenzhen decreased the size of the tea to the nanoscale to increase its bioavailability and enhance the absorption thus enhancing flavor. Such a technology could be extended to different kinds of foods like coffee, sugar, salt and spices. It could also be extended to the nutraceutical field and food supplements by decreasing the size of the delivered components to increase their effectiveness.

References

1. Frost & Sullivan (2011) Global nutraceutical industry: investing in healthy. Author, India
2. (n.d.) http://www.nutraceuticalbusinessreview.com/technical/article_page/Futureproofing_nutraceuticals_with_nanotechnology/103574. Accessed 24 Aug 2015
3. Wiwanitkit V (2012) Delivery of nutraceuticals using nanotechnology. Int J Pharm Invest 2:218
4. Huang Q, Yu H, Ru Q (2010) Bioavailability and delivery of nutraceuticals using nanotechnology. J Food Sci 75:R50–R57
5. (2010) Biotech for wellness: driving successful R&D and licensing in nutraceuticals through new business models and collaboration. Reportlinker. PR Newwire, New York
6. Tiju J, Morrison M (2006) Nanotechnology in agriculture and food. Institute Nanotechnology. European Nanotechnology Gateway
7. Frewer LJ, Norde W, Fischer A, Kampers F (eds) (2011) Nanotechnology in the agri-food sector. Wiley, Weinheim
8. Siegrist M, Cousin ME, Kastenholz H, Wiek A (2007) Public acceptance of nanotechnology foods and food packaging: the influence of affect and trust. Appetite 49:459–466
9. (n.d.) Nutralease company profile. http://www.zoominfo.com. Accessed 12 Sept 2015
10. (n.d.) http://www.haaretz.com/print-edition/business/shemen-gets-back-to-oil-1.70565. Accessed 27 Nov 2015
11. Reisch LA, Scholl G, Bietz S (2011) Better safe than sorry: consumer perceptions of and deliberations on nanotechnologies. Int J Consum Stud 36:644–654

12. (n.d.) http://www.forbes.com/2006/01/09/in_jw_0109soapboxslide.html. Accessed 27 Nov 2015
13. (n.d.) http://www.nutraingredients.com/Suppliers2/AquaNova-nanotech-extends-CoQ10-uses-bioavailabilty. Accessed 26 Nov 2015
14. Robson A (2010) Nanotechnologies and food: 1st report of session 2009–10. House of lords-science and technology committee, 2: Evidence. p 336–361
15. (n.d.) http://www.naturalproductinsider.com/news/2013/11/frutarom-aquanova-ag-market-novasol-curcumin-aspx. Accessed 23 Feb 2015
16. (n.d.) http://2020science.org/2008/10/20/shaking-up-the-nano-food-debate/#ixzz3ACitFcZV. Accessed 20 Feb 2015
17. (2010) Sustainable development commission. House of lords science and technology select committee: call for evidence; behaviour change submission
18. (n.d.) http://cciproduct.com/articles/91-nanotechnology.html. Accessed 12 Nov 2015
19. (n.d.) http://www.businesswire.com/news/home/20050329005895/en/Royal-BodyCare-Reports-Higher-Annual-Earnings. Accessed 27 Nov 2015
20. (n.d.) http://articles.chicagotribune.com/2005-05-20/business/0505200136_1_gum-venture-capitalists-chocolate. Accessed 11 Jan 2015
21. (n.d.) http://www.prnewswire.com/news-releases/olala-foods-expands-distribution-network-54407832.html. Accessed 15 Jan 2015
22. (n.d.) http://www.mrsupplement.com.au/asn-nano-creatine. Accessed 24 Jan 2015
23. (n.d.) http://www.asn-nutrition.com.au/. Accessed 26 Feb 2015

Chapter 12
Packaging Applications

Nano-based packaging represents a fast growing segment within the packaging industry due to its ability to satisfy consumer needs as well as its ability to add and enhance multiple properties within the packaging systems. Such properties include improvement of the mechanical properties, antimicrobial properties, smart detection systems of spoilage and ripeness as well as improvement in the product stability. This has resulted in the introduction of different types of nano-based packaging materials in the market including bottles, storage bags and boxes, packaging films and aluminum foil. Among the main drivers behind the increase in nano-based packaging are product enhancement and the increase in the nano-trend within the packaging industry, thus forcing different packaging companies to invest in this area of research in order to be able to compete in such a growing market segment.

Introduction

Packaging demand has witnessed a noticeable decrease due to the recent economic crisis. Consequently, this has led to an increase in the global competition among manufacturers to be able to deliver a well differentiated product that is able to satisfy the needs of the consumers. Within this context nano-based packaging is able to provide noticeable differentiation from the traditional available packaging systems [1].

Through the past decade consumers have shown a growing demand for sophisticated and higher quality packaging. However, due to the economic crisis, such high quality packaging is being demanded at affordable prices. Nano-based packaging was recognized to be able to satisfy the increasing demands of the consumers. One of the major demands of consumers is extended shelf life, an area where nano-based packaging shows multiple benefits and promising contribution [1]. Such benefits include improvement in the mechanical properties, strong barrier properties, antimicrobial properties and improvement in the stability of the packaging materials and thus the product [2].

M. Anis et al., *Nanovate*, DOI 10.1007/978-3-319-44863-3_12

Nano-based packaging involves the utilization of materials in the nanoscale with dimensions of less than 100 nm. Silver, nanoclay and titanium nitride are among the most commonly used nanomaterials (nanofillers) in the food and beverage packaging industry. Polymer nanocomposites that are used in nano-based packaging are made via dispersion of the nanofiller throughout the polymer matrix. Compared to other techniques of packaging, using polymer nanocomposites represents a cheap advanced packaging technique because it uses less expensive fillers with enhanced compressive and tensile strengths [3].

Nano-based packaging involves two sub-technologies: intelligent packaging and active packaging. Although active packaging is the dominant technology segment, intelligent packaging is growing at a faster rate. Intelligent packaging offers information regarding the freshness of the product, expiration date, the best period to be used and also offers alerts regarding the presence of microbes, gases and chemical contaminants. Thus, it offers important information to customers. Active packaging on the other hand allows the extension of the shelf life of the products via interacting chemically or biologically with the contents or head space of the product. This could be achieved via the inclusion of antimicrobial agents, oxygen scavengers, moisture absorbers or ethanol and carbon dioxide emitters. It is important to note, however, that nano-based packaging suffers from strict regulations from the US Food and Drug Administration, which recommended in 2012 a safety assessment of products that involve these types of packaging [3].

The global nano-based packaging market for both food and beverages was estimated in 2013 to be valued at 6.5 billion dollars and is expected to grow at a compound annual growth rate of 12.7 % between 2014 and 2020 reaching 15 billion dollars by 2020 [3]. The increase in export activities of different food products as well as the increase in the global demand for meat products are strong factors behind the growth of the nano-based packaging market [3]. The nano-based packaging market's major players include Bemis Company, Amcor Limited, Tetra Pak and Tetra Pak International SA, Chevron Phillips Chemical Company, Sealed Air and Klöckner Pentaplast.

Cases

SABMILLER'S BEER BOTTLES

Beer Bottles with enhanced flavor and stability.

SABMiller enhanced the stability and the flavor of its beer via the use of bottles that contain clay nanoparticles. The inclusion of the clay nanoparticles within the beer bottles noticeably decreased their permeability.

Innovation

- SABMiller introduced new beer bottles that exhibit a marked decrease in permeability for gases resulting in a better flavor for the product and increased stability.

Nanotechnology

- The company used clay nanoparticles that fill up space in the bottle's walls better than plastic molecules.

Results

- The addition of the clay nanoparticles to the bottle made it more difficult for the fizzy carbon dioxide to escape from the beer bottle and for oxygen, which could lead to spoilage of the beer, to get in.

Drivers

Strong competitive market. The nano-based packaging market is a fast growing market. Smart packaging specifically is growing faster than previously predicted.

Product enhancement and maintaining market lead. SABMiller is the second ranked beer producer according to revenues. The bottles it fabricated maintain the maximum content of carbon dioxide and prevent oxygen from entering and thus maintaining the flavor. They provide additional advantages to the company's beer products and help it maintain its position as one of the giants in beer manufacturing.

Sustainable development. SABMiller runs different sustainable development initiatives: one of which is bottles made better and more sustainable through nanotechnology.

Barriers

Safety. Public health bodies raise concerns about the migration of nanoparticles from packaging to food. Moreover, concerns about environmental contamination are also raised. Thus, legislators must consider such risks when considering approval for the widespread use of these materials.

Public acceptance. Studies indicate that consumers are hesitant to purchase nano-based products related to food. Although the concern is less in packaging than when the nano-material is included in food, still public acceptance is one of the main things that will shape the future of nanotechnology' commercialization [4].

Required process of learning. A learning process is required to educate the consumer about the added benefit of nanotechnology in the packaging material and at the same time to acknowledge the safety of this technology [5].

Enablers

Strong company and strong R&D. SABMiller dates back to 1895. Currently, it is the world's second largest brewer in the market. It has advanced R&D that enabled it to reach its leading position in the market.

Science push. The increase in nano-based research and nano-based products represents a science push for SABMiller to invest in that field.

Lean regulatory rules. SABMiller benefitted from the lean regulatory rules of the FDA. The FDA does not require nano-particles to be proved safe, but only that the product in general to not be harmful [6].

Impact

Internal. The introduction and success of the clay nanoparticles in the SABMiller bottles has encouraged SABMiller to invest more into nanotechnology and implement it more into its products. Thus, new research is ongoing with scientists at CRANN (the nanoscience institute based at Trinity College, Dublin), who have partnered with SABMiller to examine nanosheets of boron nitride to increase the shelf life of bottled beer in plastic bottles.

Partnerships. SABMiller is a major bottler for Coca-Cola and offers packaging for its products. Thus, innovative bottles will help SABMiller maintain strong partnerships as well as open new partnerships as a bottler for other strong brands.

New market shares. With its clay nanoparticle based beer bottles, SABMiller is able to maintain new market shares as a supplier of smart packaging material.

What's Next?

SABMiller added clay nanoparticles to the beer bottles which decreased greatly the gas exchange. An invention like this can be extended to different forms of food and beverage packaging.

Moreover, it can be used in boxes used for food preservation, refrigerator boxes and plastic bags.

BAYER: PACKAGING FILM

Lighter, stronger and more heat resistant packaging films.

Bayer developed packaging films that are lighter, stronger and more heat resistant. Such enhanced properties were achieved via the inclusion of silicate nanoparticles within the packaging films.

Innovation

- The packaging films developed by Bayer Polymers is stronger, more heat resistant and lighter than other packaging films currently on the market.

Nanotechnology

- Silicate nanoparticles are used to enrich the packaging films. Such new film is known as a "hybrid system" that reduces the exit of moisture and the entrance of oxygen and other gases, thus preventing the spoilage of food.

Results

- The developed packaging films prevented the packaged contents from drying out and also protected them from oxygen and moisture.

Drivers

Strong market. Smart packaging is one of the fastest growing sectors among the packaging industry.

Product enhancement. Low levels of nanoparticles are enough to noticeably change the properties of the packaging materials with no or little change in density, transparency, and processing characteristics [7].

Nano-trend. The total nano-enabled packaging market for food and beverage in 2013 was valued at 6.5 billion dollars and is expected to reach 15 billion dollars by 2020 [3].

Barriers

Safety. The rapid use of nanotechnology in packaging raised many concerns from different public health bodies regarding its safety to the consumer and the environment, especially due to lack of knowledge about the nanoparticles' possible interactions [2].

Public acceptance. Public acceptance will shape the market for nano-enabled products. Therefore, a thorough process of consumer education is required in order to increase the consumer acceptance of nano-enabled products [2].

Research and development. A certain level of infrastructure and a continuous process of research and development are required to integrate nanotechnology into the products.

Enablers

Strong company and strong R&D. With the increased trend of nanotechnology, Bayer's strong R&D was able to conduct research and develop solutions and products.

Partnerships. Partnership with Nanocor: Bayer uses Nanocor's clay in its developed packaging film to decrease the oxygen transmission rate by about 50 % [7].

Lean regulatory rules. No specific guidelines for nanoparticles were set by the FDA; no investigations are required about the fate of the nanoparticles. The sole requirements are the usual regulatory guidelines proving the safety of the product [6].

Impact

Product. The properties of the product improved drastically; the stiffness doubled and anti-blocking properties were improved. The gloss and clarity of the nano-modified product allowed it to compete with the costly high clarity polyamide film [7].

Internal: development of other nano-based products. After the introduction of Bayer's clay nanoparticle-based packaging films, the company developed additional nano-based products like the silicate nanoparticle-based hybrid system.

Competitive advantage. The improved properties of the developed packaging film gave it a competitive advantage over other similar products in the market as well as other costly high quality products.

What's Next?

Bayer's technology included clay and silicate nanoparticles in the plastic films which changed the films' properties of stiffness, clarity and gas exchange. This technology can be extended to food packaging films, plastic bottles and containers as well as refrigerator boxes and surface coatings.

MELITTA'S ALUMINUM FOIL

Aluminum foil with an enhanced ability to preserve heat.

Melitta Coffee Company has included carbon nanotubes in its aluminum foil which increased its ability to preserve heat.

Innovation

- Melitta invented an aluminum foil that exhibits an increased preservation of heat.

Nanotechnology

- Carbon nanotubes incorporation within the aluminum foil allowed the aluminum foil to reach 100°C higher surface temperatures during cooking and prevented sticking.

Results

- The technology enabled faster cooking of food with a non-stick aluminum foil.

Drivers

Strong market. At present, the world's packaging market stands for more than 3.5 billion dollars with nano-based packaging is significant and fast growing sector [7].

Incremental innovation. Melitta, which operates in the food and household sectors, by inventing a heat preserving aluminum foil, can add incremental innovation to these market sectors [2].

Nano-trend. The nano-enabled food and beverage market is among the fastest growing nano fields. Companies seeking to increase market shares should integrate nanotechnology into their plan for development and growth [3].

Barriers

Safety. Safety of the nanosized material is a critical issue due to the ability of these materials to diffuse to all parts of the body. Carbon nanotubes in particular raise certain safety issues [2].

Public acceptance. The rapid use of nano-based packaging without enough accompanying safety studies, coupled with decreased learning, is still hindering the public acceptance of nano-based products. Increasing public acceptance through a good process of learning can certainly shape the future of nano-based products [2].

Enablers

Strong R&D. Melitta has been producing bottleneck aluminum filters since World War II; it has a strong R&D that enabled it to continuously integrate new technologies.

Lean regulatory rules. Melitta group benefited from the lack of specific regulatory rules for nanomaterials where no specific safety data is required for the nanomaterial.

Strong network. The Melitta Group has production as well as administration sites in South America, Europe, North America and Asia [6].

Impact

Keeping up with a competitive market. A strong company like Melitta, in both the food and the household sector, had to catch up with the increasing nano-trend in such sectors.

Product enhancement. Low levels of carbon nanotubes increased the ability of the aluminum foil to maintain heat and allowed faster cooking.

Nano-knowledge. The nano-based knowledge, infrastructure as well as expertise used for development of the aluminum foil benefitted Melitta in many different market sectors.

What's Next?

Melitta used carbon nanotubes to develop aluminum foil that maintains heat and prevents sticking. Such technology can be extended to heating plates for food that can allow food to be heated in the oven fast, similar to the microwave but without its harmful effects. Also, it can be used in cooking bags, provided that the safety of the carbon nanotubes can be verified.

THE SHARPER IMAGE: STORAGE BAGS AND BOXES

Storage boxes that can kill bacteria

Antimicrobial storage bags and boxes were developed by The Sharper Image. The antimicrobial property was achieved through the inclusion of silver nanoparticles within the produced products.

Innovation

- Sharper Image was able to develop the miracle food-storage containers that maintain food fresher and four times longer than similar products devoid of the nano modification.

Nanotechnology

- Silver nanoparticles add antimicrobial properties to the polypropylene material included thereby reducing the growth of microorganisms. Moreover, a silicone gasket locking system keeps out the oxidizing air from entering.

Results

- Developed boxes and bags that delay spoilage of food and keep it fresher for four times longer.

Drivers

Strong market. Sharper Image is in a competitive market, the packaging market. This market is showing an increase in nano-enabled products. To maintain its shares in rapidly growing market, Sharper Image had to stay updated with current advanced technologies [7].

Competence enhancing. The inclusion of antimicrobial silver nanoparticles within its products is considered a competence enhancing innovation to its line of packaging which will allow it to maintain its original position and gain new market shares.

Nano-trend. The increase in nano-research and innovation allowed nanotechnology to be easily accessible to different companies trying to be part of such a growing trend.

Barriers

Safety. The possibility of the nanomaterial leaking from the nano-based containers represents a safety hurdle that requires scrupulous testing to ensure its safety [2].

Public acceptance. Public acceptance is one of the main barriers against the widespread use of nano-based products especially with many public organizations opposing nano-based products that have direct human contact [2].

Decreased learning. As with any new technology, a good process of learning is required to allow the widespread demand for such a new product.

Enablers

Strong R&D. Sharper Image is nationally and internationally known for its continuous supply of new and innovative high quality products. A large portion of its sales are from products developed by the company's development group.

Lean regulatory rules. The absence of specific regulatory rules for nanomaterials included in products allowed the easy approval for this innovation [6].

Strong network. Sharper Image has 190 specialty stores across the US. It also markets its exclusive proprietary via business to business sales teams for corporate incentive and reward programs as well as wholesale to select international and US retailers.

Impact

Keeping up with a competitive market. The nano-enabled food and packaging markets are among the fastest growing fields in nanotechnology. An innovative company like Sharper Image had to keep up with this trend to maintain and gain new markets [3].

Enhanced product. The added nano-innovation to its boxes will allow Sharper Image boxes to have antimicrobial properties and keep the food preserved fresher.

New market segments. Such enhanced packaging containers represent new market segments where other companies are participating in, e.g. BlueMoon Goods Fresh box that uses silver nanoparticles. Sharper Image's nano-based innovative food storage boxes allowed it to maintain its shares within such market segments.

What's Next?

Sharper image incorporated silver nanoparticles into the polymer of its food storage boxes and bags which added to its antimicrobial properties and helped maintain food fresher for longer. Such a technology can be extended to a whole range of different sectors as:

- antimicrobial textiles
- antimicrobial clothes (socks)
- antimicrobial wound dressings
- antimicrobial surfaces (in hospitals and food factories)
- antimicrobial baby toys
- antimicrobial kitchen ware (chopping boards)

RIPESENSE SMART PACKAGING

Jenkins Group's sensors are able to detect the ripeness of fruit from the packages

RipeSense has included nanosensors within its packages that allow the consumer to detect the degree of ripeness of the fruit through a color change in the package.

Innovation

- RipeSense has developed the world's first intelligent label that changes its color according to the ripeness of fruit.

Nanotechnology

- The company developed nanosensors that detect the different aromas produced from the fruit at its different ripening stages.

Results

- Market lead: first world ripe sensor of its kind. It was first introduced in New Zealand and has now reached an exclusive agreement with Fresh Technologies UK for UK distribution partnership.

Drivers

First in the market (Market lead). RipeSense is the world's first intelligent ripeness indicator label.

Strong partnership. Strong partnership between New Zealand Jenkins Group (supplier of New Zealand's horticultural labelling industry) and HortResearch (New Zealand Crown Research Institute) and their combined efforts for developing cutting edge innovation.

Nano-trend. Increased nano research and enabled technology allowed for the idea and development of such nanosensor.

Barriers

New entry. Any new entry is associated with the risk of public rejection. No matter how huge of a successful innovation it is, it can still face huge failure as there would be no previous indicators.

High tech research. Such new innovation required years of research and collaboration between different entities to develop the product. The combined effort took over 7 years of research and development to end up with this cutting edge innovation.

Decreased learning. A good process of learning was required to illustrate the competitive advantage of the developed product.

Enablers

Partnership. The partnership between Jenkins Group, a supplier to labeling industry, and HortResearch institute allowed the combined efforts of the industry and research to end up with this innovation.

Cutting edge research and development. HortResearch Institute is dedicated to cutting edge horticultural science research, thus its research, combined with the industrial need, allowed the development of the product.

Nano-enabled technology. The availability of nano-enabled technology in the past two decades allowed for the feasibility of such an innovation.

Impact

Market lead. By developing RipeSense, Jenkins Group achieved market lead with its first of a kind intelligent labeling.

Enhanced product. The RipeSense intelligent labeling greatly enhanced the properties of the labeling creating unprecedented levels of efficiency.

New markets. RipeSense, the first in the market with its enhanced labeling properties, extended from its country of invention, New Zealand, to other markets like the UK through an exclusive agreement with Fresh Technologies UK.

Market Dominance. As the first and the only in the market with such intelligent packaging, the company achieved market dominance in this segment.

What's Next?

The RipeSense label developed by Jenkins Group in partnership with HortResearch institute depends on nanosensors that can detect the ripening of a pear depending on the produced aroma (volatile gases). Such a technology can be extended to all forms of food sensing, not only ripening, but also to detect food spoilage.

References

1. (2013) Global nano packaging market 2013–2023; opportunities for nanotechnology. Visiongain, KD4R6, London
2. Market Publishers (2009) Nano-enabled packaging for the food and beverage industry: a technology, industry and market analysis. Innovative research & products N2701020A76EN
3. Persistence Market Research (2014) Global market study on nano-enabled packaging for food and beverages: intelligent packaging to witness highest growth by 2020. Persistence Market Research, New York
4. Frewer LJ, Norde W, Fischer A, Kampers F (eds) (2011) Nanotechnology in the agri-food sector. Wiley, Weinheim
5. (2004) Down on the farm: the impact of nano-scale technologies on food and agriculture. ETC group
6. (n.d.)http://www.scientificamerican.com/article/do-nanoparticles-in-food-pose--health-risk/. Accessed 23 Nov 2015
7. Boye JI, Arcand Y (2012) Green technologies in food production and processing. Springer, New York

Chapter 13
Agricultural Applications

Nanotechnology has been recently introduced to the agriculture sector with a few goals in mind. One of which is to introduce enhanced fertilizers with less loss in nutrients and pesticides and with decreased side effects to the environment and the surrounding tissues. Moreover, nano-based sensors could allow management and delivery of water, pesticides and fertilizers as well as soil analysis. Despite the different sectors of the agriculture industry that nanotechnology could have an impact on, nano-based agriculture products have weak presence in the market. This is attributed to the high cost of production as well as public concerns about the safety of using nano-based products in a sector that is directly connected to their daily consumption of food. Thus, further research is needed to decrease the costs of production as well as to establish sufficient safety studies concerning agricultural products.

Introduction

The application of nanotechnology in the agriculture industrial sector aims to achieve different outputs: reduction of the usage of harmful plant pesticides, reduction in loss of nutrients during fertilization and increasing the production yield of different crops. Different nano-based structures benefit the agricultural field. Examples of these structures include nanocapsules and nanoparticles, which are used in the treatment and detection of diseases, proper delivery of nutrients to specific sites and in enhancing the plant's absorption of nutrients. Specific nanoparticles are able to reduce damage to the environment and to the healthy tissues of the plant. Within this context, nanoparticles derived from biopolymers, like proteins and carbohydrates, hold special interest within the agriculture sector as they are relatively safe to the human health and the environment. Starch-based nanoparticles are examples of carbohydrate-based nanoparticles that have been extensively investigated as sustainable and nontoxic delivery systems for agrochemicals and biostimulants.

M. Anis et al., *Nanovate*, DOI 10.1007/978-3-319-44863-3_13

Nanosensors are one of nanotechnology's contribution to the agriculture sector. Such nanosensors can be used for soil analysis, management and delivery of water, easy biochemical sensing and control and delivery of nutrients and pesticides. Carbon nanotubes, nanofibers and fullerenes have been recently involved in sensitive biochemical sensors. Recently, agricultural waste products have been considered for the production of fossil fuels and nanomaterials.

Although nanotechnology holds great potential in the agricultural sector, the nano-based agriculture market is still comparably marginal compared to other fields where nano-based products have already witnessed a strong market presence. Some major issues are still to be resolved in order for the nano-based agriculture market to show realizable growth. Increasing the production scale while decreasing the cost stands as one of the significant issues facing nano-based agricultural products. This is especially important in the agriculture sector because the high cost of production of nano-based products is easily compensated in other sectors like the medical and pharmaceutical sectors while this is not the case in the agricultural sector.

Cases

SYNGENTA PESTICIDE

Pesticides with decreased side effects.

Syngenta has utilized nanoemulsions within its developed pesticides. This allowed controlled delivery of the pesticide in response to certain stimulus. Thus, it increased the pesticides' benefits and decreased the side effects.

Innovation

- Syngenta developed nano-pesticides that allow controlled delivery of pesticides to increase their benefit and decrease their side effects.

Nanotechnology

- Nanoemulsions that encapsulate different pesticides and allow their controlled delivery according to different stimulus.

Results

- A line of different products has been developed using the nanoemulsion technology; a plant growth regulator: a pesticide used for different crops that only breaks open upon contact with leaves.

Drivers

Strong competitive market. Nano-based pesticides are already on the market. Major agrochemical firms like BASF and Bayer have patents for nano-biocides. The companies have active R&Ds for developing nanoscale pesticide formulations.

Product differentiation/nano-trend. The strong competition with companies like BASF and Bayer and the increased research and the growing trend of nano-based technologies are considered major drivers. Syngenta used its strong scientific foundation to increase its market shares through the development of nano-based products.

Population growth and scarcity of land. Population growth and scarcity of land is one of the main drivers in the agrochemical sector for developing new innovative solutions [1].

Barriers

Safety. Safety is a major concern with nano-based applications related to food. Nano-pesticides decreasing the size of the pesticide raises safety concerns for both the farmer, the environment and the plants consumed by animals and humans.

Public acceptance. Public acceptance is a major hurdle in the face of any food-related nano-based application since safety is not completely ensured through long term studies [2].

Enablers

Strong scientific foundation. Syngenta's strong scientific foundation allowed it to develop the nano-emulsion technology [1].

Science push. Increase in nano-related research as well as the associated technology and expertise represented a science push towards developing nano-based pesticides.

No regulatory hurdles. A known pesticide that is newly formulated in a nano-form does not require any regulatory re-examination since it is not considered as a new chemical [2].

Impact

Internal. The infrastructure required for the nano-based product is extended to other products produced by the company giving it the lead in nano-based agrochemicals.

Industry: competitors moving in the same path. The introduced nano-based product by Syngenta has encouraged major players in the agro-industry to conduct R&D on nano-based products.

Market lead. Syngenta, being the first to introduce a nano-based pesticide, has achieved a lead within this sector.

What's Next?

Syngenta has developed nanoemulsions that allow controlled delivery of the pesticides in response to certain stimulus, thus increasing the efficiency and decreasing the side effects. Such technology could be extended to drug delivery systems and to other pesticides. It could also be extended to insecticides applied in homes so as to increase their effect against the insects.

SONIC ESSENTIALS™

Nano-fertilizer for rapid growth!

New fertilizer developed by Sonic using nanotechnology has found success in the optimum delivery of nutrients to plants for long durations of time.

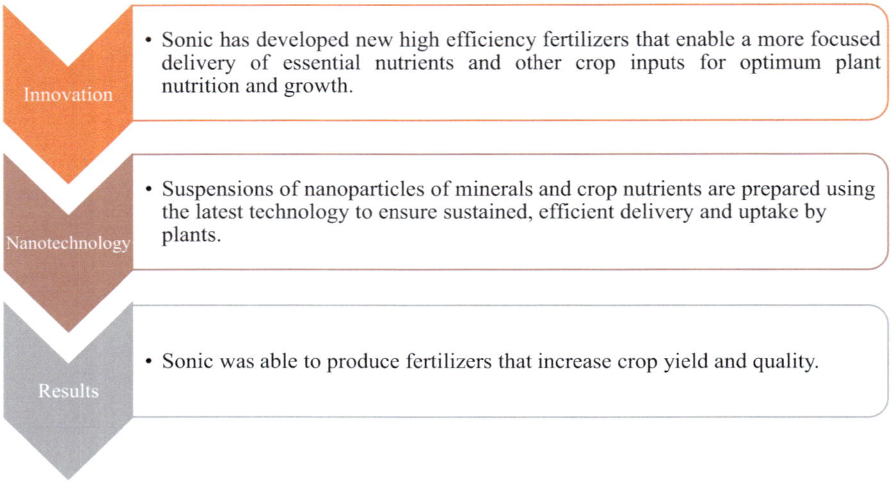

Drivers

Increasing world population and food demand. According to a report by the High Level Expert Forum (2009), by 2050, the world population is expected to increase to 2.3 billion people. It is also expected that the per capita incomes will increase as well. This means that there will be an increase in demand for food. Demand on cereal, for both food and animal feed uses, is predicted to reach some three billion tons by 2050. This means that the industry of agriculture has to increase its crop yield to keep up with this rapidly growing demand.

Increased demand for good quality food. The global economy is projected to grow by 2.9 % annually until 2050 which means that people will be able to spend more on high quality foods and, therefore, demand for high quality food will significantly increase.

Need to avoid economic losses by farmers. The death of crops or low production yields causes farmers to lose the money, time and effort spent on such plants.

Barriers

Technology. The technology of nanomaterials and the equipment needed for achieving such small size is costly. However, due to Sonic's long career in this field, it accumulated the facilities and experience needed as well as a good staff of chemists, engineers and biologists which helped it overcome this barrier.

Safety issues. The use of nanomaterials in fertilizers has been constantly subject to scrutiny due to its possible leakage to underground waters or plants sold, posing possible dangerous effects on the human health upon ingestion. This requires strict regulations and licensing to ensure enough tests have been conducted regarding toxicity.

Enablers

Experience. Sonic Essentials has been working in the field of fertilizers since 2007 which enabled it to conduct a great amount of research, lab testing of the fertilizers and analysis of the market response.

Certification and licensing. Sonic's nanosized products are licensed by the Therapeutic Goods Administration and the Australian Pesticides and Veterinary Medicine Authority.

Impact

Due to the quality of the fertilizers, the company was able to sell them in Australia and globally to Asia, Africa and the Middle East.

What's Next?

The next step for Sonic is to use its technology to make targeted fertilizers in which the nanoparticles can be attached to linkers that will anchor them to the roots of plants.

References

1. Shirley I (2007) Controlled delivery of agrochemicals. Presented at the Chemspeceurope Symposium, RAI Amsterdam
2. (2004) Down on the farm: the impact of nano-scale technologies on food and agriculture. ETC group

Chapter 14
Cosmetic Applications

Nanocosmetics represent a fast growing segment within the cosmetic industry. They are cosmetic products with nano-based structures included either as nano-inclusions or as delivery vehicles. Nanocosmetics exhibit enhanced properties compared to traditional cosmetics including improved skin hydration, improved skin penetration, enhanced stability and bioavailability. Major cosmetic companies like L'Oréal, Christian Dior and Estee Lauder are investing in nanocosmetic research as well as nano-patents. Different nanocosmetics have already made their way to the market in the form of sunscreens, makeup, lotions and shampoos. Enhanced properties of the nano-based products and keeping up with the nano-trend are among the main drivers behind the increase in nano-based cosmetics. Yet the safety of the nanocosmetics is one of the main barriers against its wide application. One of the main enablers, however, that allowed the introduction of large number of nanocosmetics till now is the absence of specific regulatory rules concerning nano-based cosmetics.

Introduction

Nanotechnology is strongly forcing its way into the cosmetics industry in the form of nanocosmetics. The dynamic presence of nanotechnology in the cosmetics industry is due to the enhanced properties imparted by nanotechnology on different cosmetics including enhancement in performance, color, stability and transparency [1].

Nanocosmetics has found its way in different products including sunscreens, moisturizers, makeup and hair care. Nanotechnology is utilized in the cosmetic products either as nanoparticle inclusions or as delivery vehicles. The former is used in cosmetics like UV filters. The major compounds used in this approach include zinc oxide and titanium oxide as well as some organic alternatives. In the case of delivery vehicles, niosomes and liposomes are mainly used. However, solid lipid nanoparticles and nano-structured lipid carriers have been shown to be better delivery vehicles and can provide better stability,

M. Anis et al., *Nanovate*, DOI 10.1007/978-3-319-44863-3_14

skin hydration and bioavailability. Nanoemulsions and nanocrystals as encapsulation techniques have also been investigated for applications in the cosmetic industry [2].

Almost all major cosmetic manufacturers use nanotechnology in their products. L'Oréal is directing 600 million dollars of its revenues to nano-patents [1]. However, recently concerns about the safety of the use of nano-ingredients in different cosmetics has been raised [1]. Notably, the legal requirements for nanocosmetics are the same as those required for ordinary cosmetics. This has facilitated the approval of nanocosmetics in products and allowed their increased presence in the market. Recently, however, there is an increasing demand that standard safety tests should be developed for nanoingredient-containing cosmetics in order to validate their safety and overcome the public debate [2].

Cases

CHRISTIAN DIOR FOUNDATION

Waterproof foundation that lasts longer.

Cosmetics giant Christian Dior introduces a waterproof, long lasting foundation through the use of Nano-Stretch Network™ technology.

Innovation

- Diorskin nano-based foundation cream is characterized by a waterproof formula that is able to last for a long time with no transfer even in extreme conditions.

Nanotechnology

- Patented Nano-Stretch Network™, inspired by nano-textile technology, creates a mesh of foundation on the skin that is micro-aerated and invisible for a perfect finish and color. It also contains a hydragel system that allows moisture stabilization for optimal comfort, hour after hour.

Results

- Diorskin Extreme Fit Wear with its nano-stretch network allowed extreme fit of makeup, layered as if it is a second skin.

Drivers

Catching up with the nano-trend. L'Oréal, one of Christian Dior's competitors, has begun its nanotech research in the 1980s and has many nano-based products in the market. Thus, to catch up with such an increasing trend and prevent loss of market shares, Dior was forced to search and invest in nanotechnology and research.

Nanotechnology and research. Nanotechnology research has witnessed a massive advancement providing nano-enabled innovations for companies who seek to utilize it.

Improved products. Nanotechnology in cosmetics has gained increased importance due to the enhanced properties added via the integration of nanoparticles compared to other conventional technologies employed in cosmetics. Such properties include enhanced solubility, color, transparency and UV protection [3].

Barriers

Safety. Due to the nanosize, these nano-entities are able to penetrate natural biological barriers like the digestive, alveolar-capillary and blood brain barriers. As a result, sufficient safety studies should be conducted to validate their use [4].

Public debate. Due to controversial studies about the safety of nano-based cosmetics, the label nano on the product might be viewed skeptically by some consumers while be appealing to others [3, 4].

Enablers

Strong R&D. Since 1986, Christian Dior's strong R&D allowed it to be the first to incorporate liposomes in its cosmetics. Ever since then, Christian Dior has continued that line of research enabling the making of different nano-based products.

Enabled technology. As it started its liposome manufacture in 1986, Christian Dior has developed the enabling technology that allowed it to be innovative in the field of nanotechnology and easily apply its inventions.

Regulatory gap. The FDA's regulations are the most lenient in the cosmetics sectors [5].

Impact

Internal; production of different nano-products. The successful application of nanotechnology in its products has enabled Dior to extend the application of nanotechnology to more of its products [6].

Keeping up with a market trend. Nanotechnology in cosmetics is one of the fastest growing fields. Thus, integration of nanotechnology allowed Christian Dior to keep up with such growing trend [7].

Maintaining its market shares. L'Oréal, one of Christian Dior's major competitors ranks, sixth among nanotechnology patent holders in the US. Thus, integrating nanotechnology allowed Dior to keep up with the market with its nano-enhanced products thereby avoiding loss of market shares.

What's Next?

As the Nano-Stretch Network™ technology utilized by Christian Dior was inspired by the nano-textile technology, it could be used in the development of advanced wound dressings with micro aerated mesh. Such technology could also be extended to sunscreens, face creams and other cosmetics as did Christian Dior.

SHINSO SKIN LIFTING GEL

The miracle skin lifting gel; looking younger is now achievable.

Shinso Essence utilized nano-fullerenes to develop a nano-based gel with maximum antioxidant activity and maximum penetration to the skin, thus allowing an incredible skin lifting effect.

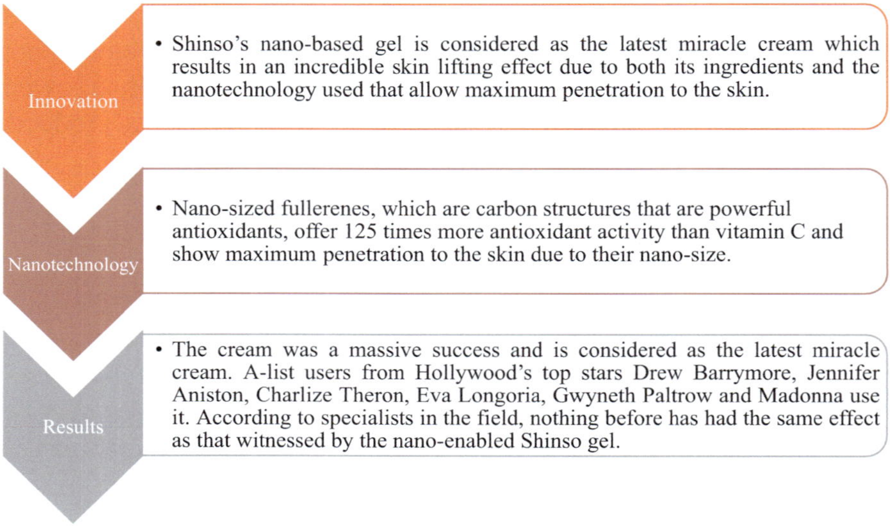

Drivers

Catching up with the nano-trend. Nanotechnology is an area that all of the leading cosmetic companies are quickly turning to. Such companies include L'Oréal, Procter & Gamble, Estée Lauder, Shiseido and Dior. Thus, to be a major player in the market and compete with the leading companies, nano-enabled products are a must to be sought.

Expertise of the inventor. Shinso made use of Tsuruta's expertise, an aerospace engineer and the inventor of this technology, who then turned to designing deep water desalination systems. He used seawater that was 2000 ft. deep and merged it into other products, one of which is fullerene nanotubes, and came out with his miracle product [8].

Enhanced nano-related properties. Fullerene carbon nanostructures are more powerful antioxidants than vitamin C. The antioxidant property is further strengthened by the nanosize, which also enhances penetration between skin cells into the dermis [8].

Barriers

Safety. Alarmists of the nanotechnology field warn about the ability of nanoparticles to penetrate the skin barrier and enter the bloodstream. This requires efficient and extensive safety studies to be carried out [7].

Public debate. There is an ongoing and increasing public debate regarding the potential hazards of nanotechnology on the human health and the environment [7].

Good learning process. A good learning process is required so as to make the consumer aware of the added benefits of nanotechnology in the product coupled with safety data about the product.

Enablers

Available expertise. Tsuruta's expertise was the main enabler for the product. He utilized his knowledge of the properties of deep-sea water from his past experience in deep water distillation systems coupled with his knowledge as an engineer in utilizing fullerene nanostructures to come up with this superior product [8].

Sold technology. The nano-fullerenes used in the product were purchased from a trusted company that conducted the safety studies on the utilized fullerenes [8].

Registered as cosmeceuticals. Shinso's nano-based gel has a physiological effect on the skin. Such products ought to be registered as drugs. However, the company, to avoid the difficult regulatory hurdles of being registered as a drug, registered it as a cosmeceutical. This helped avoid the time and money-consuming regulatory procedures associated with drug registration [8].

Impact

Market lead among world class actresses. Due to the miracle anti-ageing effects of the Shinso nano-based gel, it showed market lead among Hollywood stars.

Keeping up with a market trend. Nanotechnology in cosmetics is one of the fastest growing fields and the integration of nanotechnology allowed Shinso to keep up with such a growing trend [7].

Consumer enhanced quality. A conversation between a beauty journalist and Tsuruta went as follows: "I'm used to seeing the skincare world throw up one of these wonder products every month, but I've never seen anything do that to a face. How, I ask him, does it work?" asked the beauty journalist. "Lifting in itself is very difficult to do with natural ingredients," Tsuruta, the inventor, says. "What makes Shinso different is the blend of ingredients and the way it penetrates the skin" [8].

What's Next?

Nano-Fullerenes are powerful antioxidants, thus they could be extended to advanced wound dressings, skin therapeutics as well as other cosmetics like creams and lotions. However, proper safety studies should always be conducted.

PUREOLOGY SHAMPOO AND CONDITIONER

Shampoo that transforms hair.

Pureology introduced a shampoo that allows incredible improvement of aged and colored hair via the use of nanoparticles that were 15–20 nm in size.

Innovation

- Pureology introduced its nano-enabled shampoo, a product that allowed the dramatic improvement of aged and color-treated hair, thereby enhancing its color and shine.

Nanotechnology

- Nanoparticles that were 15–20 nm in size (about 1/57,000th the human hair size) improved the product's ability to moisturize and strengthen the hair as well as preserve its color and provide thermal protection. This can be attributed to improved penetration, availability and distribution of the material at the nanosize.

Results

- The success of its nano-enabled product allowed Pureology to integrate nanotechnology in all of its shampoos, reconstructors, conditioners, and moisturizers. Moreover, Pureology recently launched a new line that contains nano-ingredients that fill the hair with moisture and strength and help it resist humidity more efficiently.

Drivers

Demand pull. There is a growing demand for hair that is shinier and healthier and, to some extent, the available shampoos still do not satisfy that need.

Nano-trend. Due to the increased trend of making nano-based cosmetic products with enhanced properties, Pureology began trials for incorporating nanotechnology into its products from a performance point of view [9].

Trials with promising results. After testing the inclusion of nanotechnology into its products, the company's trials revealed that, by decreasing the particle size to 15–20 nm, the ability to strengthen and moisturize the hair was improved drastically [9].

Barriers

Safety. Ability of the nanoparticles to penetrate different body barriers and penetrate deep into the skin raises safety concerns among consumers [7].

Public Debate. Controversial studies have raised public debate about using nanotechnology and its effects on the environment and human health. These still represent barriers against using nano-enabled products, especially those that are to be in direct contact with the human body [7].

Enablers

Curious R&D. The R&D in Pureology was curious and willing to improve, despite its already high quality products, to try integrating nanotechnology and testing the change in the performance of its products.

Lean regulatory rules for nanocosmetics. There is a regulatory gap in the FDA in analyzing nano-based products, especially cosmetics and food supplements; there is no special scheme developed by the FDA for nano-based products [5].

Acquisition by L'Oréal as part of its professional product division. In 2001, Pureology used nanotechnology for its line of nano-based hair products. Six years later, the company was acquired by L'Oréal as part of its professional product division, which contributes to more than 14 % of the total revenues of the company [10].

Impact

Internal. The company developed its one of a kind successful line of products. Markham, CEO and Founder of Pureology, says "Pureology now uses nanotechnology in all of its shampoos, conditioners, moisturizers and reconstructors. The company recently launched its super luxury category of products all using nanotechnology" [9].

Consumer: enhanced quality. The nanotechnology in Pureology products showed improvements for curly, hair which tends to be dry. The nano ingredients fill the hair with moisture and strength and allow it to better resist humidity.

Keeping up with a market trend. Nanotechnology in cosmetics is one of the fastest growing fields. Thus, by integrating nanotechnology in its products, Pureology was able to keep up with such a growing trend [7].

What's Next?

Pureology decreased the size of its ingredients to 15–20 nm. At such a nano size, the material exhibits different properties including enhanced solubility, greater surface area, a reduced amount of ingredients, improved distribution and bioavailability. This draws attention to the ability with which such a technique can be integrated into a whole range of different products like other cosmetics, the food and beverage industry and the pharmaceutical industry.

KARA VITA LIP TENDER

Makeup with maximum skin penetration and enhanced effect.

Kara Vita introduced a lip tender with enhanced potency due to its high skin penetration via the use of nanosphere encapsulating ingredients.

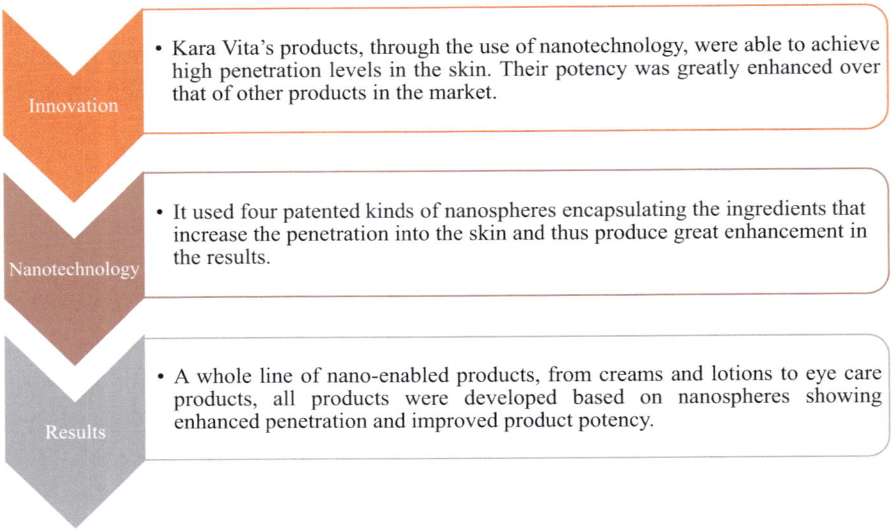

Drivers

Unsatisfied with the results of products. When Kara Vita was established, it noticed that none of the already existing products produced satisfactory results. The main problem was that, whatever superior ingredients were used, the product always seemed to remain at the surface [11].

Nano-trend. Nearly all the major companies in cosmetics are investing heavily in nanotechnology and, besides nanotechnology's added benefits in products, Kara Vita was encouraged to invest in it with the aim of achieving excellence and market lead.

Achieving market lead and product excellence. Kara Vita and its founder aimed to achieve the best of skincare, market lead and product excellence. Therefore, he had to integrate nanotechnology to overcome problems in penetration in order to maximize the effect.

Barriers

Safety. Although decreasing the particle size of the ingredients increased the penetration of Kara Vita's products, it also increased penetration to different body barriers like the blood brain barrier and digestive barrier. As a result, the safety of the products must be thoroughly investigated [7].

Public debate. Despite the added benefit of nanotechnology in different products, controversial studies have raised public concern and debates where certain groups, like Friends of the Earth International, are totally against nanotechnology, especially in human related nanoproducts [7].

Enablers

Ambitious founder. According to the company Founder and Chairman Karalyn Schuchert, "It all began after a long search to discover the best skin care product. It did not matter what the ingredients advertised, or how expensive and exotic they were, none of them provided the penetration and hydration I had hoped for. The product always seemed to remain on the surface of my skin!" [11].

Acquiring a nanotechnology company. The opportunity for acquiring a company known for its revolutionary nanotechnology has enabled the development of the product.

Strong team. Not only did the company integrate nanotechnology, but, due to problems of raising the products to pharmaceutical standards, the company also collaborated with Deborah Duffey, one of the most creative product developers. This has resulted in a line of superior cosmetic products.

Patented technology. The nanospheres developed by Kara Vita were protected and patented to ensure the market lead of the company with its patented technology.

Impact

Internal. The technology has enabled the establishment of a line of successful products. Kara Vita now has a line of different products with its patented nanotechnology.

Product excellence and superiority. The nanotechnology utilized by Kara Vita does not only improve penetration and thus better moisturize and nourishment to the skin, but also helps repair and strengthen the skin. With the company's patented Lyphazome nanotechnology after depleting the loaded ingredients, the phospholipid nanospheres form a matrix with the surrounding cells which helps in rebuilding the skin's lipid layer.

Keeping up with a market trend. Integrating nanotechnology into Kara Vita's products allowed it to keep up with one of the fastest growing trends in cosmetics; nanotechnology [7].

What's Next?

The nanospheres utilized by Kara Vita not only deliver the ingredients with improved penetration, but also help to rebuild and strengthen the skin through the phospholipid composition. Thus, such nanospheres can be applied in all different cosmetic products as was accomplished by Kara Vita. It can also be integrated with excellent results in advanced wound dressings and in skin care pharmaceutical products.

SANGI'S TOOTHPASTE

Toothpaste that can actually repair damaged teeth.

Sangi introduced the first toothpaste that is actually able to repair the damaged tooth enamel via the use of nano-hydroxyapatite "mHAP" remineralizing toothpaste.

Innovation

- Japan's Sangi introduced the first nano-hydroxyapatite "mHAP" remineralizing toothpaste that exceeds the normal functions of toothpaste in actually having the ability to repair the damage of the tooth enamel.

Nanotechnology

- The technology uses nanocrystals of hydroxyapatite, which is a main component of the tooth enamel. During brushing, "mHAP" binds with proteins and adheres to plaque fragments and bacteria, facilitating their removal. This property is greatly enhanced by the "mHAP" nanoparticle size which leads to an increase in the surface area for protein attachment. The hydroxyapatite nanocrystals remineralize demineralized areas of the enamel by supplementing the mineral supply from saliva. It also helps restore mineral density and translucency in order to enhance teeth's whiteness.

Results

- Sangi has sold more than 90 million tubes of toothpaste and continues to expand its line of products containing nanoparticles.

Drivers

Technology transfer. NASA had proposed hydroxyapatite for restoring the minerals lost from the teeth of astronauts in a gravity free environment. Sangi conceived the idea and launched the first hydroxyapatite toothpaste [12].

Market lead. By launching the first enamel restorative toothpaste, Sangi achieved market lead in this sector [12].

Competitive advantage. No other toothpaste in the market contained the natural tooth component nano-hydroxyapatite, thus the ability of the company to have developed such a component gave it a competitive advantage over other competitors in the market.

Science push. The developed synthetic hydroxyapatite by NASA represented a science push towards developing the nano-hydroxyapatite toothpaste.

Barriers

Intensive Research. To develop the nano-hydroxyapatite toothpaste Sangi spent over a decade in laboratory research and field trials [12].

Optimization of the size. Sangi continued upgrading the developed technology and optimizing the size of the nano-hydroxyapatite for maximum benefit. In 2003, it succeeded in decreasing the size of the nano-hydroxyapatite nanoparticles for maximum results [12].

Enablers

Patenting. Sangi acquired different patents for its nano-hydroxyapatite technology. Early lab work and field trials were carried out since the 1980s during which Sangi was devoted to intensive research and field trials on the enamel restorative properties of the nano-hydroxyapatite. In 2003, Sangi had optimized the nanosize of its nano-hydroxyapatite particles [12].

Collaboration. Tokyo Medical and Dental University and Gifu Dental University (now Asahi University) both collaborated with Sangi in its research. Moreover, the universities conducted field trials proving the enamel restorative power of the nano-hydroxyapatite toothpaste [12].

Regulatory approval. The Japanese government in 1993 approved Sangi's proprietary form of hydroxyapatite. It was designated as a Medical Hydroxyapatite to differentiate it from other hydroxyapatites in other dental applications [12].

Impact

Internal: establishment of a line of products. Sangi's nano-hydroxyapatite is used now in toothpastes and remineralizing chewing gum. In addition, other applications with the same technology are under development in collaboration with Tohoku University.

Market lead. One of major enamel restorative toothpastes by Sangi was introduced in 1985 and for 30 years has been sold extensively in Japan with over 90 million tubes sold.

Competitors following the same path. BASF and Henkel, noticing Sangi's success, announced the development of nano-hydroxyapatite in 2002. Henkel then launched the first nano-hydroxyapatite toothpaste.

Product excellence. The product that was first introduced by Sangi showed excellent results in occluding exposed dentinal tubules and reducing hypersensitivity as well as exhibiting enamel restorative properties.

What's Next?

Other applications that are now under development by Sangi include hydroxyapatite Powder Jet Deposition system (PJD) for the treatment of tooth decay. In addition, home-care anti-caries systems were also developed. Thus, such a technology can be applied to all tooth related care products due to its enamel restorative property as well as other beneficial effects. It can also enter into bone related products and therapeutics due to the benefits of nano-hydroxyapatite for bones.

References

1. Raj S, Jose S, Sumod US, Sabitha M (2012) Nanotechnology in cosmetics: opportunities and challenges. J Pharm BioAllied Sci 4:186
2. (n.d.) Nanotechnology in cosmetics. http://www.nanowerk.com/nanotechnology-in-cosmetics.php. Accessed 21 July 2015
3. Nasir A (2009) "Nano present" and "nano future": the growing role of shrinking technology in dermatology, Part 1. Cosmet Dermatol 22:139
4. Wu X (2012) Nanotechnology in cosmetics: a review. Cosm Toil 127(6):266–272
5. Fender J (2008) The FDA and nano: big problems with tiny technology. Chicago-Kent Law Rev 83:1063
6. Dana DA (ed) (2011) The nanotechnology challenge: creating legal institutions for uncertain risks. Cambridge University Press, Cambridge
7. (n.d.) http://www.forbes.com/2006/05/03/ge-motorola-jnj-in_jw_0504soapbox_inl.html. Accessed 14 June 2015

8. (n.d.) https://shinsoskincare.com/pressdetail/thetimes_fab11.php. Accessed 28 June 2015
9. (n.d.) http://www.naturallycurly.com/curlreading/curl-products/size-matters-the-nanotech-explosion/. Accessed 19 July 2015
10. (n.d.) http://www.bidnessetc.com/business/nanocosmetics-harmonizing-science-and-beauty/. Accessed 12 Aug 2015
11. (n.d.) http://www.karavita.com/aboutus/foundersmessage.aspx. Accessed 11 Apr 2015
12. (n.d.) http://www.apagard.com/en/web/history/index.html. Accessed 29 Aug 2015

Chapter 15
Pharmaceutical Applications

A large number of already existing drugs suffer from poor water solubility, burst release and unfavorable side effects due to un-targeted delivery. Implantation of nanotechnology in the drug industry allows overcoming these problems and introducing drugs with enhanced water solubility, sustained release and targeted delivery to the diseased site. Moreover, nanotechnology allows reformulation of several hundred billion dollars of unused drugs instead of spending billions of dollars in formulating new drugs. Currently, several nano-based drug formulations have made their way into the market. Science push, observed side effects of the original product and developing better formulations are among the major drivers behind developing nano-based drugs. Meanwhile, safety concerns, regulatory approvals and required expertise and infrastructure are among the major barriers against the introduction of new nano-based drug formulations.

Introduction

The drug industry is witnessing a major shift away from the one drug fits all approach to drugs that are more personalized based on the patient's' unique immune response and genome. Such shift to more personalized medicine will make use of targeted nanotechnology-based drug delivery systems [1]. Moreover, a large number of existing drugs suffer from poor solubility. The application of nanotechnology in the drug industry will allow benefit from unused, several hundred billion dollars' worth of already existing drug compounds [2]. Nano-enabled drug delivery systems will result in extending the patent life-cycles of different drugs via reformulating the existing drugs and/or compounds in new and more effectively delivered formulations [2].

Generally, many drugs suffer from poor water solubility that subsequently results in low dissolution rates and thus weak activity. Other drugs suffer from burst release that necessitate frequent administration of the drug. Untargeted delivery of different drugs with unnecessary and associated side effects represent an additional problem in

M. Anis et al., *Nanovate*, DOI 10.1007/978-3-319-44863-3_15

drug formulations. Decreasing the size of the drug and/or compounds to the nano range increases the surface-to-volume ratio and increases the dissolution rates. Nano-based drug delivery systems allow introducing drug formulations that are able to overcome solubility problems, are targeted to the site of the disease and are able to achieve sustained release over prolonged periods of time.

According to the latest market reports, the total nano-enabled drug delivery market size is forecast at 136 billion dollars by 2021 with nanocrystals and nanocarriers as the most commonly used nano-based structures. Still, developing novel drug delivery systems could create more value for companies and entrepreneurs [1]. In a nanocarrier drug delivery system the drug is loaded into/on the nanocarrier (also known as nanoparticles or nanoshells, between 1 and 100 nm) that could be labeled with targeting moieties allowing targeted delivery to the target site. Among the different nanocarriers available, gold nanocarriers and liposomes represent 45 % of the total market. Within the next decade liposomes will hold the largest market that could reach 15 billion in 2021. Meanwhile, gold nanocarriers will witness the highest compound annual growth rate of 53.8 % [1].

A schematic illustration of the added effects of nano-based drug formulations.

Cases

PFIZER'S IMMUNOSUPPRESSANT TABLETS

Increasing the solubility of the drug allowed formulating it in tablet form.

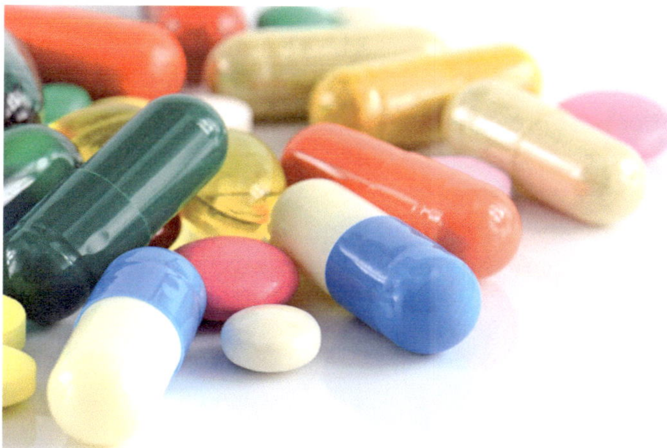

Pfizer has reformulated an immunosuppressant drug in the nanocrystal form. The original drug was poorly soluble and could only be formulated in the powder form that had to be reconstituted before administration. Yet upon formulating it in the nanocrystal form, its solubility was greatly enhanced and that enabled formulating it in the tablet form.

Innovation

- An immunosuppressant for renal transplant patients is a poorly soluble drug. Pfizer formulated it in the nanoform. This has resulted in enhancement of its solubility and thus overcoming the poor dissolution rate and poor pharmacokinetics.

Nanotechnology

- Nanocrystals: the drug is formulated as oral tablets of nanocrystals of the active constituent sirolimus. Such nanoformulation led to an increase in the surface-to-volume ratio, thus when compared to a micronized formulation of the same drug it had better solubility, better bioavailability and better dissolution of the drug.

Results

- The success witnessed with the nano-based immunosuppressant drug has been extended to three other approved drugs to enhance their solubilities. Whereas before the application of nanotechnology the drug could only be formulated in powder form that had to be reconstituted before administration, after the application of nanotechnology, the drug could be formulated as a tablet.

Drivers

Drawbacks of the original product. The low water solubility of the products did not only decrease its bioavailability and distribution and thus decrease its medicinal effect, but also required the administration of higher doses of the drug [3].

Science push. The increased research and development in the nanotechnology, field especially in drug delivery and pharmaceuticals, represents a great science push to the pharmaceutical industry [3].

Better formulation. The poor water solubility of the drug caused it to be only available as a powder dosage form that required a complicated reconstitution procedure for its administration [3].

Barriers

Cost-benefit ratio. Major European countries' markets are cost regulated concerning the drugs. This is a significant limiting factor towards developing innovative drugs that have higher costs [4].

Safety concerns. Many of the properties of the materials are changed when they are decreased to the nano-range. Such new properties should be tested for their efficacy, especially that they could diffuse through different barriers causing many unintended side effects [5].

Expertise and Infrastructure. The new innovation required expertise as well as infrastructure in the field of nanotechnology.

Enablers

Technological and research advancements. Technology as well as scientific advancement in the field of nanotechnology allowed the development and large scale production of the nanocrystals which in turn allowed the increased solubility of the drug [3].

Regulatory modifications. The FDA speeded up the drug regulatory process and provided guidance in the form of Biopharmaceutics Classification System (BCS). Such system identifies expendable clinical bioequivalence tests [6].

Extended benefit. The costs spent on new infrastructure and research in that area could be compensated by applying this developed technology into different products of the company that suffer from the same problem; low solubility.

Impact

Internal. The company acquired new expertise and infrastructure that was extended to a number of different products.

Market lead and increased profits. The immunosuppressant drug that was based on the nanocrystal technology became the fastest selling drug within the transplant market. In 2008, the sales of the drug reached 376 million dollars.

Industry. The nanocrystal technology is being used in four other drugs [3].

Patient. Enhanced patient convenience with the new tablet formula compared to the previously troubling reconstituted powder formula [3].

What's Next?

Pfizer uses wet ball milling to decrease the particle size of the active constituents, thus increasing the surface-to-volume ratio. This in turn increases the bioavailability, solubility, dissolution and reactivity. Such technology can be extended to different pharmaceutical formulas suffering from low solubility or reactivity. Moreover, such technology can be extended to the food sector, where decreasing the particle size of preservatives and flavors will enhance their effects and also decrease the added concentration of such materials.

ANTICANCER DRUG BY ALZA, JOHNSON & JOHNSON

Delivery of an anticancer drug led to decreased side effects.

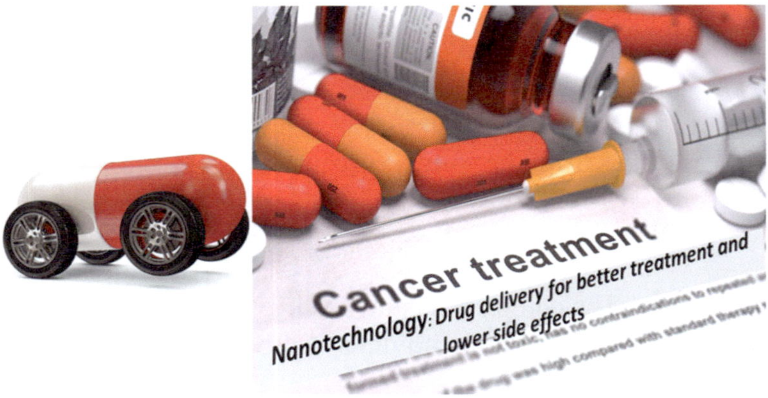

ALZA utilized nanotechnology to develop a new anticancer drug that treats ovarian cancer and Kaposi's sarcoma. The utilization of nanotechnology allowed the drug to escape the immune system and deliver the precise dose of the drug to diseased areas.

Innovation

- A new anticancer treatment that can deliver the precise dose of the drug to diseased specific areas by escaping from the immune system. The drug treats ovarian cancer and Kaposi's sarcoma.

Nanotechnology

- Lipid nanoparticles that have a Polyethylene Glycol (PEG) coating (STEALTH® technology).

Results

- The developed STEALTH technology helps to escape the impact of the immune system and allows precise delivery of drugs thus increasing the efficacy of treatment, decreasing the dose and the associated side effects.

Drivers

Increased side effects. About 90% of the non-modified drug was lost in the body which necessitated increasing the dose and led to the unfavorable side effects of cancer treatments [7].

Science push. The increase in nano-based research represented a science push towards such innovation [7].

Competitive advantage. Current cancer treatments mostly all suffer from the same drawbacks of increased side effects and unspecificity. Thus, ALZA, with the incremental innovation of its anticancer drug, could acquire new market shares from its competitors.

Barriers

Regulatory approval. As ALZA's anticancer drug was the first lipid based drug to be introduced in the US and EU markets, it faced conservative and cautious assessment from the regulatory bodies [8].

Efficacy and safety assessment. Long studies for validating the efficacy and the safety of the newly introduced lipid-based drug were required.

Infrastructure and expertise. The new STEALTH technology required acquiring new expertise and new infrastructure by the company.

Enablers

High resolution characterization. ALZA used high resolution characterization methods (qNano) to verify the particle size within the definition of bioequivalence as set forth by the USFDA and EMA so it could ensure the optimization of the new formulation [8].

Decreased cost of incremental innovation. The nano-based form of an already existing drug had low research and development costs compared to developing a completely new drug.

Increase in nano-research. There is an increase in nano-based research, especially in drug delivery, where it is considered the dominant research field contributing 76 % of the scientific publications in nanomedicine [7].

Impact

Internal. ALZA developed excellent expertise in handling the STEALTH technology that can be extended to a whole range of similar drugs.

Competitors. The increased efficacy and decreased side effects of the nano-modified anticancer drug provided it with a competitive advantage over other competitors in the market.

Patient. Compliance and acceptability by the patients due to the decreased side effects.

What's Next?

The developed STEALTH technology by ALZA allowed the drug to evade the immune system and enhanced its delivery to the required diseased parts. This technology can be extended to similar cases of drugs with increased side effects in different diseases not only in cancer drugs.

NOVAVAX TOPICAL LOTION FOR MENOPAUSE

Estradiol lotion instead of orally administered drug.

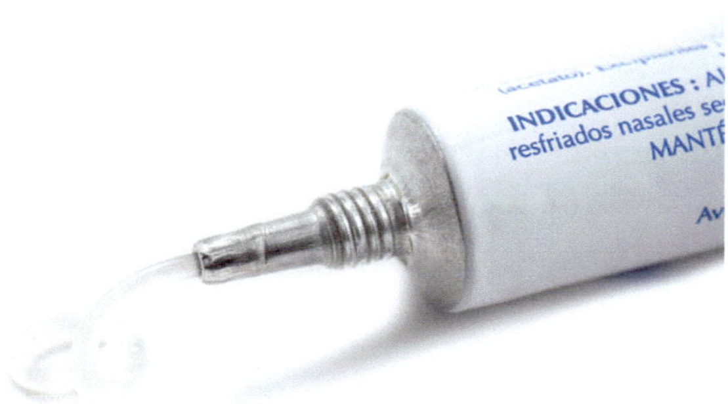

Estradiol that is used in the treatment of menopause was usually administered orally and subjected to the clearing effects of the liver. Novavax utilized nanotechnology to develop the first estradiol topical lotion which increased patient convenience and enhanced the effectiveness of the drug.

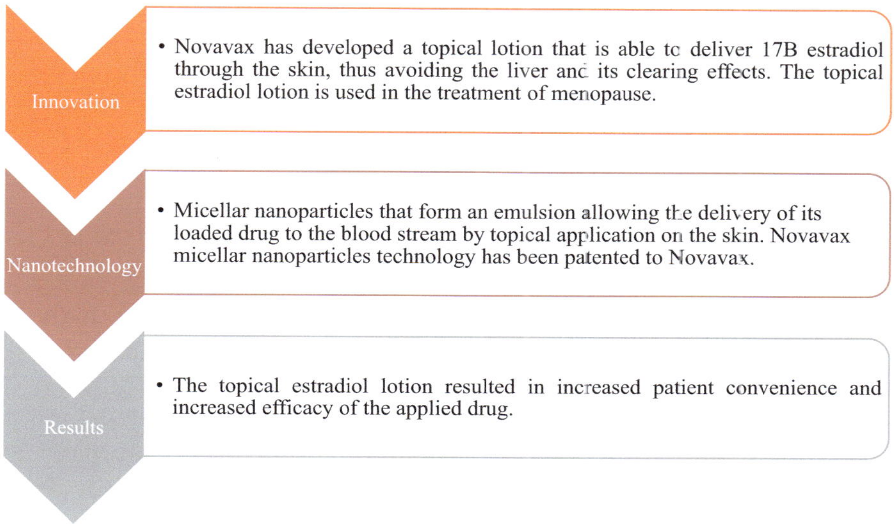

Drivers

Nano-trend and research. There is a growing trend of conducting research in nanotechnology based formulations. Moreover, there are always great and novel advancements in nanotechnology research.

Strong market. The developed product would compete in the non-systemic segment of the Hormone Replacement Therapies (HRT) US market, which is the only segment showing growth within the 1.5 billion dollar US HRT market [9].

Internal strategy. Novavax is a company focused on creating differentiated and value-added pharmaceutical, vaccine products and technologies.

Improved delivery route; decreased drawbacks. The new delivery route could overcome the side effects associated with the original delivery route because the new route bypasses the liver.

Barriers

Research and development. Intensive research, optimization and clinical studies were required to develop the nano-based drug that allowed administration of the drug in an optimal dose via topical application.

Intellectual Property (IP). Novavax had to acquire IP for its developed technology in order to maintain its competitive advantage in the market.

Enablers

Strong experience. Novavax utilized similar drug delivery technologies in other products. Such technologies include Novasomes® and Sterisomes® for subcutaneous injection.

Company internal structure. Novavax works in developing and producing vaccines. Thus, the R&D in the company is familiar with the development of micellar systems for drug delivery and for proper vaccine delivery.

Impact

Licensing. The improved properties due to the acquired and patented technology by Novavax had Esprit Pharma license exclusive North American rights from Novavax [9].

Partnerships. The licensing of the topical estradiol lotion by Esprit was the beginning of a long term relationship to expand the product pipeline of Esprit [9].

Patient. According to research studies, the topical estradiol lotion is widely accepted by patients due to its better route of administration [9].

What's Next?

Novavax used micellar nanoparticles to prepare a preferable topical dosage form of its drug where the micellar nanoparticles enhanced the delivery of the drug via the skin thus allowing it to bypass the liver. Such technology could be extended to other drugs allowing a preferable topical dosage form of these drugs. Also, such a technology could be extended to cosmetics, creams and lotions increasing the diffusion and transfer of its components via the skin.

CELGENE'S ANTICANCER DRUG

Overcoming the insolubility of an anticancer drug.

Celgene utilized nanotechnology to overcome the insolubility of the main active constituent of a breast cancer drug. The developed nano-based breast cancer drug improved activity with marked decrease in side effects.

Innovation

- A breast anticancer drug (FDA approved in 2005) that could overcome the insolubility of its active constituent paclitaxel.

Nanotechnology

- Nanoparticles (approximately 130 nm) of albumin bound to the active constituent paclitaxel.

Results

- The breast anticancer drug total revenues from 2009 were 314.5 million dollars. The stock of the company soared after its approval by the FDA making the company's chairman (who owns over 80 % of the company's stock) a multi-billionaire. Moreover, the application of nanotechnology allowed the patients to overcome the hypersensitivity reactions that were usually associated with the unmodified drug due to the solvents used for increasing the solubility in the unmodified formulations, and thus increasing the patient's convenience and acceptability.

Drivers

Science push. The marked increase in nano-research in the past two decades, especially in drug delivery with a special focus on cancer, represented a science push towards enhancing such an important class of therapeutics.

Strong market. Breast cancer is the most common cancer in women. Its incidence increases with age, doubling nearly every 10 years until menopause, as provided by Globocan (2002). Thus, breast cancer drugs have a strong market.

Demand pull. Paclitaxel is a strong anticancer drug, however, it suffers from insolubility. This necessitates the use of solvents that usually lead to hypersensitivity reactions in the patients. Therefore, an enhanced drug delivery system was required.

Barriers

Intensive research and clinical trials. Developing an optimized new formulation of the drug requires extensive research and expensive clinical trials.

Regulatory approval. FDA approval of new drugs always represents a long and expensive process that stands in the way of new drug development.

Enablers

Previous patent expiry. The original version of the nanoformulation is Taxol. Upon the expiry of Bristol-Myers Squibb's Taxol patent in 2000, a generic version of it, Paclitaxel, became available at about 150 dollars per dose [10].

Proper use of nano-coating. The superiority of the developed breast cancer drug is due to its innovative albumin coating. Albumin is a natural protein in the body that has few allergic reactions that were often associated with paclitaxel as severe side effects [10].

Patent monopoly. Nanotech-enabled drugs allow extending and securing monopoly patents by making new nano-versions of existing old patents of drugs [10].

Impact

Market acquisition. Abraxis BioScience was the first to develop the Nanoparticle Albumin Bound (NAB) technology platform. Such technology was then acquired by Celgene which now markets the developed nano-based anticancer drug.

Strong market and multi-billionaire CEO. The developed nano-based anticancer drug total revenues in 2009 were 314.5 million dollars. The stock of the company increased after the drug's approval by the FDA making the company's chairman a multi-billionaire.

Patient. Enhanced patient compliance due to decreased side effects.

Nanodrugs. The approval of the developed nano-based anticancer drug by the FDA was considered a giant leap for nano-enabled drugs [10].

What's Next?

Celgene used albumin (natural protein in the body) based nanoparticles to decrease the allergic reactions produced by the body's immune response to foreign bodies or drugs. Thus, such a technology can be extended to other drugs decreasing the allergic reactions associated by them and decreasing their side effects.

SMITH & NEPHEW'S ANTIMICROBIAL WOUND DRESSING

Wound dressings that kill bacteria.

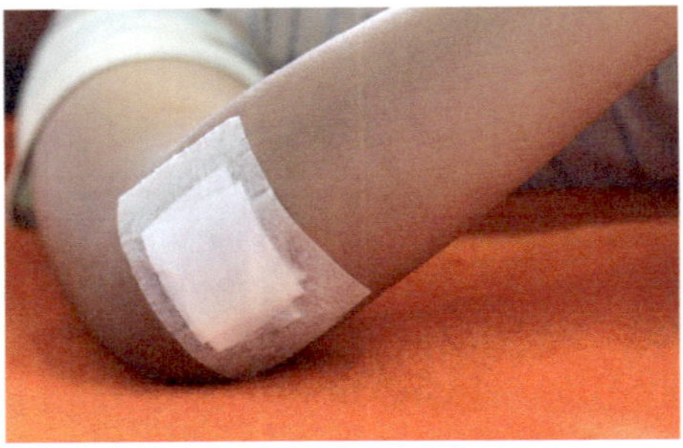

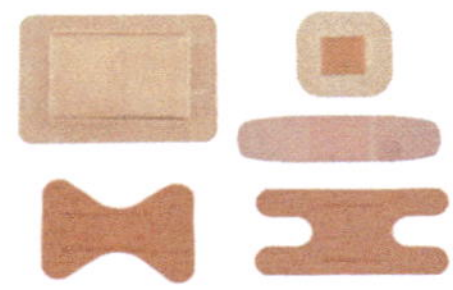

Smith & Nephew developed wound dressings loaded with nano-crystalline silver. The nano-crystalline silver imparted enhanced antimicrobial properties to the developed dressing.

Innovation

- A wound dressing with enhanced ability for fighting infections.

Nanotechnology

- Silcryst nano-crystalline silver that impart enhanced antimicrobial properties to the dressing.

Results

- Effective wound dressing for reducing risks of infections associated with wounds. Moreover, the nano-enabled wound dressing allowed the Smith & Nephew silver based dressings to enter a new market segment; Infection Management.

Drivers

Competence enhancing. Smith & Nephew silver based dressings is a significant development to the dressings developed by the company, thus improving activity in infection management [11].

Demand pull. Smith & Nephew's silver based dressings were developed after the company listened to clinicians' main concerns and noticed their need for infection management wound dressings [11].

Nanosilver trend and research. There is increased dedication to nano-research with a massive portion of it directed towards silver nanoparticles and their associated properties, especially their antimicrobial properties.

Share in a new market segment. The Smith & Nephew silver based dressings allowed the company to focus on a new market segment of wound care: infection management [11].

Barriers

Intensive research and clinical trials. Research and clinical trials are needed to develop efficient and optimized wound dressings coupled with studies to validate their safety.

Nanosilver safety debate. Despite the wide use of silver nanoparticles, their safety is still not completely validated. Several studies have linked negative side effects to silver nanoparticles. Thus, public debate about the use of silver nanoparticles is among the barriers standing against it.

Enablers

Partnership. The development of Smith & Nephew's silver based dressings was possible due to its partnership with Nucryst Pharmaceuticals, the developer of the Silcryst technology [11].

Network. Nucryst licensed this technology to Smith & Nephew, which has had strong market presence for over 30 countries. Moreover, Smith & Nephew prides itself with its strong network of professional health care customers and surgeons with whom the company's name is well known for.

Strong R&D. Smith & Nephew is one of the leading companies in advanced wound management, endoscopy and orthopedic trauma. Thus, it has well positioned R&D that ensures continuous development of high-tech products.

Impact

New market shares. Smith & Nephew's silver based dressings allowed the company to have shares in new markets, which in this case was infection management in wound therapy.

Product excellence. The Smith & Nephew silver based dressings have been designed after listening to physicians and professionals in the field and developed the optimized products that would support them in wound infection management. The Smith & Nephew silver based dressings' line combines both enhanced antimicrobial properties (nanocrystalline silver) and enhanced conformability on awkward anatomical areas such as hands and face. That's in addition to the excellent stretch capabilities of the dressing.

What's Next?

Smith & Nephew used nanocrystalline silver to impart antimicrobial properties to its wound dressing. Thus, such a technology of antimicrobial silver nanoparticles could be extended to any application that would benefit from having antimicrobial properties. Examples of different applications could be socks, where antibacterial silver based socks would prevent foul odor, bed dressings in hospitals to decrease nosocomial infections and fridges.

References

1. Cientifica (2012) Nanotechnology for drug delivery: global market for NanoCarriers. Author, London
2. Cientifica (2007) The nanotech revolution in drug delivery. Author, London
3. Junghanns JUA, Müller RH (2008) Nanocrystal technology, drug delivery and clinical applications. Int J Nanomed 3:295

4. Von der Schulenburg F, Vandoros S, Kanavos P (2011) The effects of drug market regulation on pharmaceutical prices in Europe: overview and evidence from the market of ACE inhibitors. Health Econ Rev 1:1–8
5. De Jong WH, Borm PJ (2008) Drug delivery and nanoparticles: applications and hazards. Int J Nanomed 3:133
6. Kunde DS (2015) Biopharmaceutical classification system: a brief account. Int J Res Meth 1(1):20–46
7. (2014) Transparency market research. Transdermal drug delivery market—global industry analysis, size, share, growth, trends and forecast. Transparency market research, New York
8. Accessed from http://www.azayatherapeutics.com/
9. Business wire (2005) Esprit pharma acquires exclusive North American rights to ESTRASORB(R) topical emulsion; sees growing pre-eminence in women's healthcare. ESPRIT PHARMA, North America
10. ETC group (2006) Nano-drug's dirty little secret. Author, London
11. Accessed from http://www.nanowerk.com/news/newsid=8089.php

Final Conclusive Words

Challenges to consider while using nanotechnology
As previously mentioned, materials behave differently at the nanoscale. Although this is the power of the nanoscale materials, it does not go without risks. Concerns rise about the toxicities of some nanomaterials because, as materials become smaller, there is fear that they might more easily cross the blood brain barrier (a membrane that protects the brain from any harmful materials in the blood stream). Thus, with the widespread use of nanomaterial in nearly everything, from antimicrobial socks and highways to walls and metro stations, precise and intense investigations regarding the long term safety of the nanoparticles must be considered.

There are also some ethical and social issues concerning nanotechnology. Fear of the misuse of nanotechnology in the creation of weapons of mass destruction raises the need for certain rules and regulations to govern the development of nano-based devices. Uncertainty concerning market and consumer acceptance is also a challenge facing the entry of nano-based products. Organizations, like Friends of the Earth, are already alarming the public about the use of nano-based products. Thus, to overcome this barrier, proper awareness regarding the properties and safety of the developed product should be available.

How did they make it?
By analyzing the different industrial sectors covered in this book along with the different case studies in each sector, we can conclude several factors that contribute to the success of companies, enterprises or governments that are willing to invest or apply nanotechnology innovations:

1. Smart Investment
The government's investment in nanotechnology is a crucial factor for businesses to flourish. Many governments have worked on developing nano-research and commercializing nano-based tech-

M. Anis et al., *Nanovate*, DOI 10.1007/978-3-319-44863-3

nologies. For example, the US government adopted investments in nanotechnology by establishing the National Nanotechnology Initiative (NNI), making nanotechnology a federal initiative. However, governments and investors should be extremely cautious when spending cash on immature technologies because there are no generalizable standards to evaluate nanotechnology developments. Additionally, it is difficult to predict which innovations would make it to the market and, for those that do, which would thrive in the market. Many innovators rushed into the market with high-tech innovations without ensuring that they can adequately develop and improve their products.

2. Open Innovation

In order to benefit from a nanotechnology innovation, enterprises do not need to study nanotechnology from scratch nor to completely overturn the scientific activities of their R&D labs—they do not even need to have an R&D lab. All what an enterprise needs is to identify a problem or an intended application. Then it can collaborate or partner with a nanotechnology company or a research center that can fulfill its expected outcome. This model is what many enterprises adopted in our discussed case studies and was proven successful. Open innovation is a big advantage to small nanotechnology companies because they can make use of the well-established R&D and manufacturing capabilities of the bigger corporations. Many have taken the adopted nanotechnology advantage to the next stage by applying it to different products, therefore, have targeted new segments.

3. Competitive advantage

The "know-how" is the secret word for building a strong business in nanotechnology-based applications. Companies that developed and secured a know-how for a particular process or product innovation managed to make growing profits from this knowledge, especially when this know-how is hard to imitate, thus reducing threats of new entrants. This explains why most of the nanotechnology companies are spin offs from successful academic research that could be up-scaled and could serve new or existing markets.

4. Educating the public

As mentioned previously, public understanding of nanotechnology is still somewhat immature. Here, regulatory authorities play an important role in educating the public about nanotechnology and in informing the public about the safety issues associated with the different types of nanomaterials. Companies also have a responsibility towards their nano-based innovations. They should attempt strategies to educate their customers and to show that they are complying with regulations to reassure their customers. This will lead to the differentiation of their innovations in a competitive market.

5. Business sense

A genius in nano is not enough. Many academic researchers or even nanotechnology companies work on innovations that are highly promising. However, they lack the talent to transform the technology into a profitable business. Sometimes they pour all their efforts and money into technology aspects and end up with not enough left to invest in creating business models or in networking with business professionals and investors.

6. Market demand

No matter how innovative the technology being introduced to the market, there should be a growing need in the market for this technology otherwise it is rendered a failure. A big market size will guarantee a profitable position for the new innovation and the innovators could then work solely on differentiating themselves. Usually nano-based innovations are new market entrants, therefore, they need to spot a growing industrial trend and use it as an incubator where the technology will not only make profit, but will also be developed, improved and sustain its path of success. Incumbents are usually big barriers to the entry of new technologies. Nano-tech based innovations should introduce differentiating features such as lower prices or better performances or both.

MIX
Papier aus verantwortungsvollen Quellen
Paper from responsible sources
FSC® C105338

FSC
www.fsc.org

If you have any concerns about our products,
you can contact us on
ProductSafety@springernature.com

In case Publisher is established outside the EU,
the EU authorized representative is:
Springer Nature Customer Service Center GmbH
Europaplatz 3, 69115 Heidelberg, Germany

Printed by Libri Plureos GmbH
in Hamburg, Germany